The Relation of Metabolism and Myoelectrical Activity in Human Skeletal Muscle investigated by Simultaneous ^{31}P Nuclear Magnetic Resonance Spectroscopy and Surface Electromyography

Peter Vestergaard-Poulsen

Ph.D. Thesis

Danish Research Centre of Magnetic Resonance, Hvidovre Hospital
Department of Medical Informatics and Image Analysis, Aalborg University

The Relation of Metabolism and Myoelectrical Activity in Human Skeletal Muscle investigated by Simultaneous ^{31}P Nuclear Magnetic Resonance spectroscopy and Surface Electromyography

ISBN87-7307-534-5

Publisher Aalborg University Press

Distribution Aalborg University Press
Badehusvej 16
DK-9000 Aalborg
Tel. +45 98 13 09 15
Fax: +45 98 13 49 15

Layout Peter Vestergaard-Poulsen

Print Schølin Grafisk

Den offentlige Ph.D. forelæsning blev afholdt på Hvidovre Hospital
d. 27. april, 1995.

CONTENTS

PREFACE

This study was carried out at the Danish Research Centre of Magnetic Resonance, Hvidovre Hospital, Hvidovre, Denmark. The study has been financially supported by: The Danish Research Centre of Magnetic Resonance, the Danish Research Academy, Aalborg University, The Danish Medical Research Council (grant no. 12-96356) and The Danish National Research Foundation.

This thesis is based on the following studies:

I. Vestergaard-Poulsen, P., Thomsen, C., Sinkjær, T., Stubgaard, M., Rosenfalck, A. and Henriksen, O. Simultaneous electromyography and ^{31}P nuclear magnetic resonance spectroscopy - with application to muscle fatigue. Electroencephalography and Clinical Neurophysiology 85: 402-411, 1992.

II. Vestergaard-Poulsen, P., Thomsen, C., Sinkjær, T. and Henriksen, O. Simultaneous ^{31}P NMR spectroscopy and EMG in exercising and recovering human skeletal muscle: technical aspects. Magnetic Resonance in Medicine 31: 93-102, 1994.

III. Vestergaard-Poulsen, P., Thomsen, C., Sinkjær, T. and Henriksen, O. Simultaneous ^{31}P NMR Spectroscopy and EMG in Exercising and Recovering Human Skeletal Muscle - A Correlation Study. Journal of Applied Physiology 79(5): 1469-1478, 1995.

IV. Vestergaard-Poulsen, P., Thomsen, C., Nørregaard, J., Bülow, P., Sinkjær, T. and Henriksen, O. ^{31}P NMR Spectroscopy and EMG during exercise and recovery in patients with fibromyalgia. Journal of Rheumatology 22: 1544-1551, 1995.

V. Nørregaard, J., Bülow, P., Vestergaard-Poulsen, P., Thomsen, C. and Danneskiold-samsøe, B. Muscle strength, voluntary activation and cross-sectional muscle area in patients with fibromyalgia. British Journal of Rheumatology 34: 925-931, 1995.

The studies are referred to by their Roman numerals.

SUMMARY

The myoelectrical changes in exercising skeletal muscle have been studied extensively, and a number of theories regarding the relationship to the underlying metabolic changes have been postulated. A better understanding of the myoelectrical changes during exercise would be gained if the relationship between the myoelectrical activity and energy metabolism were known.

The present thesis describes an investigation of the association between myoelectrical changes and energy metabolism in skeletal muscle, which was examined by combining ^{31}P nuclear magnetic resonance spectroscopy and surface electromyography in studies of the anterior tibial muscle used as a human skeletal muscle model during exercise and recovery. The myoelectrical-metabolic relationship was investigated during static exercise of submaximal intensity in healthy volunteers, a patient with McArdle's syndrome and patients with fibromyalgia.

The study indicates that during static exercise the motor unit recruitment increases hyperbolically as the oxidative potential decreases. During exhaustive exercise the motor recruitment reaches a maximum at a specific oxidative potential, which seems to constitute a consistent metabolic limit to a voluntary muscle contraction. Abruptly increasing myoelectrical activity occurs at a certain metabolic and pH level and seems to be related to the recruitment of large glycolytic motor units. The results indicate that neither pH reduction nor lactate accumulation is primarily responsible for the myoelectrical spectral changes as previously assumed. The increased myoelectrical activity seen after long duration exercise or exhaustive exercise is not caused by an incomplete recovery of phosphorous metabolism or pH, but probably by an impairment of the excitation-contraction coupling.

Patients with fibromyalgia have reduced maximum muscular force in relation to muscle area. The myoelectrical-metabolic relation is normal but the results indicate that the lower exercise tolerance found in these patients could be connected with the low physical activity levels, and may be due to intolerance of glycolytic activity.

Simultaneous ^{31}P nuclear magnetic resonance spectroscopy and surface electromyography in combination with nuclear magnetic imaging offers useful information about the energy metabolism, motor unit recruitment, morphology and spatial activation in human skeletal muscle, and could be an important research tool in non-invasive studies of basal muscle physiology, sports medicine, occupational physiology and in clinical applications.

DANSK RESUMÉ (Danish summary)

Mange undersøgelser har vist at der sker myoelektriske ændringer i arbejdende skeletmuskulatur, og har ofte udmundet i teorier om relationen til de underliggende stofskifteforandringer i musklen. En bedre forståelse af de myoelektriske ændringer under arbejde ville kunne opnås hvis relationen mellem myoelektrisk aktivitet og muskelstofskiftet var kendt. Nærværende studium beskriver hvordan de myoelektriske forandringer er relateret til muskelstofskiftet i skeletmuskulatur under arbejde og restitution, undersøgt ved simultan ^{31}P nuklear magnetisk resonans spektroskopi og overflade-elektromyografi med m. tibialis anterior som human muskelmodel. Relationen mellem de myoelektriske ændringer og muskelstofskiftet blev undersøgt under submaximalt statisk arbejde hos normale, hos en patient med McArdle's syndrom og hos patienter med fibromyalgia.

Studiet har vist, at under submaximalt statisk arbejde stiger rekrutteringen af motoriske enheder hyperbolisk i relation til et faldende oxydativt potentiale i musklen. Maximal observeret rekruttering af motoriske enheder under udtrættende arbejde er forbundet med et specifikt oxydativt potentiale, og udgør sandsynligvis en konsistent grænse af stofskiftet under en frivillig muskelkontraktion. Den pludselige ændring i den myoelektriske aktivitet under udtrættende arbejde sker ved et bestemt stofskifte-niveau og pH-værdi, og skyldes sandsynligvis rekruttering af større glykolytiske motoriske enheder. Hverken pH-værdi eller laktat ophobning har et primært ansvar for de spektrale ændringer i det myoelektriske signal, som tidligere antaget. Den forhøjede myoelektriske aktivitet, der observeres efter langtids-kontraktioner og udmattende arbejde, skyldes ikke en ukomplet restitution af fosformetabolismen eller pH, men er sandsynligvis relateret til forandringer i excitations-kontraktions-koblingen.

Patienter med fibromyalgi har reduceret muskelkraft i forhold til muskelstørrelse. Relationen mellem de myoelektriske ændringer og stofskifte-forandringerne er normale, men patienternes lave arbejdstolerance ser ud til at være forbundet med lav fysisk aktivitetsniveau og kan skyldes en intolerance overfor glykolytisk aktivitet.

Simultan ^{31}P nuklear magnetisk resonans spektroskopi og overflade elektromyografi, kombineret med nuklear magnetisk resonans billeddannelse, giver værdifulde oplysninger omkring muskelstofskiftet, rekruttering af motoriske enheder, muskelmorfologi og den spatiale aktivation i human skeletmuskulatur; og bedømmes til at være en vigtigt undersøgelsesmetode i non-invasive studier af basal muskelfysiologi, idrætsmedicin, arbejdsfysiologi og i kliniske applikationer.

ACKNOWLEDGEMENTS

I would like to thank all the staff at the Danish Research Centre of Magnetic Resonance for their support. I thank Steen Ahlmann, Peter Borch, Pia Olsen, Helle Simonsen, Nina Hansen and Anne-Marie Vind for their assistance during the experiments with patients and volunteers, and Jens Arnth Jensen, Anders Stensgaard Larsen and Simon Topp for their technical assistance, as well as Leif Dalbjerg (Siemens Medical Systems, Denmark).

Many thanks to Dr. Ron de Beer (Technical University, Delft, Netherlands) for a very fine personal introduction to Processing of Nuclear Magnetic Resonance signals, and for supplying the VARPRO software for this.

Thanks also to Maeve Drewsen for revising the language in the manuscripts.

For their ceaseless support and scientific guidance throughout the study, I wish to thank my supervisors: Research Council Professor Thomas Sinkjær, Ph.D. (Centre for Sensory-Motor Interaction, Aalborg University, Aalborg) and Professor Ole Henriksen, Ph.D. (head of the Danish Research Centre of Magnetic Resonance).

Most of all, I want to acknowledge my colleague Carsten Thomsen, M.D. to whom I am greatly indebted for the tremendous support he has given me through the entire study.

Finally, I want to thank my wife Anette for her encouragement through the years.

DEFINITIONS AND ABBREVIATIONS

Adiabatic radio frequency pulse: A radio frequency pulse that excites the spins of a sample volume equally.

ADP: adenosinediphosphate

ATP: adenosinetriphosphate

Autoregressive modelling: modelling of a signal using an all-pole signal description. Autoregressive modelling is closely related to linear prediction methods. (see e.g.: Makhoul,J. Linear prediction: tutorial review. Proc. IEEE, 63:561-580, 1975; Kay, S.M and Marple, S.L. Spectrum analysis - a modern perspective. Proc. IEEE 69:1380-1419, 1981.)

B_0: Strong static magnetic field

Chemical shift:: The total static field B_0 experienced by the spins can vary slightly due to the chemical environment, and produces slightly different resonance frequencies. The relative differences in resonance frequency are called the chemical shift.

EMG: electromyogram or electromyography. The myoelectrical activity appears from a depolarization of a muscle fibre sarcolemma.

Excitation-contraction coupling: The process by which depolarization of the muscle fibres initiates a contraction.

Fast-twitch muscle fibres: Type IIa, IIb. Muscle fibres with a short twitch duration, and high glycolytic capacity. Type IIa is an intermediate fibre with both high oxidative and high glycolytic capacity.

FFT: Fast Fourier transformation. (Cooley, J.W and Tukey, J.W. An algorithm for machine calculation of complex Fourier series. Math. Computation 19: 297-301, 1965.)

FID: free induction decay. During relaxation it is possible to detect magnetic field changes originating from the spinning nuclei with an electrical coil. The induced voltage is called the free induction decay (FID), and consists of one or more exponential decaying sinusoids with frequencies near the resonance frequency.

j-coupling: see spin-spin coupling.

McArdle's syndrome: myophosphorylase deficiency

Motor unit: A single motor neuron and the muscle fibres innervated by this neuron constitute a motor unit.

MPF: Median power frequency.
The frequency that divides the SEMG power spectrum P(f) into two bandwidths with equal integrated power:

$$\int_0^{MPF} P(f)\,df = \int_{MPF}^{\infty} P(f)\,df = \frac{1}{2}\int_0^{\infty} P(f)\,df$$

MRI: magnetic resonance imaging.

MVC: Maximum voluntary contraction. The maximum voluntary contraction force of a muscle.

MVC-10: Exercise protocol using 10% of the maximal voluntary contraction force (MVC) in a sustained static muscle contraction. The protocol also included brief contractions (10% MVC) in the recovery period.

NMR: nuclear magnetic resonance

NMR spectrum: The amplitudes of the FID are proportional to the number of nuclei contributing to the signal, and the different resonance frequencies are associated with the chemical environment of the nuclei. A frequency analysis of the NMR signal by the FFT results in a frequency spectrum presented as the frequency offsets relative to the excitation frequency. The integrated intensities of the frequency components are proportional to the concentration of molecules contributing to the FID. Only nuclei with high mobility contribute, because immobility of the nuclei within a molecule reduces the NMR visibility (an excellent review is given in: Abragam, A. "principles of nuclear magnetism", Oxford science publication, Clarendon Press, 1983.

^{31}P: phosphorous isotope with atom-mass 31

P_i: Inorganic phosphate

PCr: Phosphocreatine

PDE: Phosphodiesters

PME: Phosphomonoesters

Proton decoupling: by exciting the ^{1}H-spins by a radio frequency pulse before the ^{31}P-spins the effect of the coupling (see spin-spin coupling) between the ^{1}H and ^{31}P nuclei is reduced. The ^{31}P signal is hereby enhanced because the ^{1}H-^{31}P coupling produces a broadening of the resonances in the ^{31}P NMR spectrum.

Radio frequency pulse: the external oscillating magnetic field used for excitation (see resonance frequency)

Relaxation: Following excitation with the resonance frequency the energy state of spinning nuclei will return to equilibrium again after some time. This phenomenon is called relaxation. During relaxation it is possible to detect magnetic field changes originating from the spinning nuclei with an electrical coil (see T_1 and T_2).

Resonance frequency: When nuclei with the spin property are placed in a static magnetic field (B_0) energy can be transferred to the spinning nuclei by an external oscillating magnetic field with a specific frequency -the resonance frequency. The resonance frequency (ω_0) is dependent on the strength of static magnetic field and an isotope-specific constant γ - the gyromagnetic ratio. The so called Larmor equation links the resonance frequency and the field strength: $\omega_0 = \gamma \cdot B_0$.

RIEMG: Rectified and integrated electromyogram.

$$\frac{1}{T}\int_0^T |S(t)|dt$$

RMS: Root-mean-square.
The square-root of the mean squared signal amplitude (calculated in a time period T)

$$\sqrt{\frac{1}{T}\int_0^T S(t)^2 dt}$$

Sarcolemma: cell membrane of a muscle fibre.

Sarcoplasmatic reticulum: A tubular system surrounding the myofibrils of a muscle fibre. The sarcoplasmatic reticulum is concerned with Ca^{++} release and uptake during muscle contraction and relaxation.

Sech/tanh radio frequency pulse: A radio frequency pulse with hyperbolic secant amplitude-modulation and hyperbolic tangent frequency-modulation.

Shim: adjustment of the static magnetic field B_0.

Signal-to-noise ratio: The signal amplitude to peak-to-peak noise ratio

Sin/cos: A radio frequency pulse with sine amplitude-modulation and cosine frequency-modulation.

Slow-twitch muscle fibres: Type I. Muscle fibres with high twitch duration, and high oxidative capacity.

Spin: Certain nuclei like ^1H and ^{31}P possess a so-called precession due to the rotation (spin) about their own axis and have certain magnetic properties. The nuclei are sometimes called spins.

Spin-spin coupling: interaction between adjacent nuclei producing spectral components with a multiplet of resonances.

SEMG: Surface electromyogram or surface electromyography. The myoelectrical activity from the currently active motor units appearing at the skin surface. The SEMG is usually recorded with surface electrodes attached to the skin.

T_1,T_2: The relaxation process is characterized by two time constants T_1 and T_2. The time constant of the nuclei returning to equilibrium is called T_1. The T_2 relates to the energy exchange between nuclei and affects the decay rate of the FID.

1. INTRODUCTION

EXERCISING MUSCLE

A voluntary muscle contraction can be considered as a result of three processes: 1) neural drive from the central nervous system which conducts nerve potentials from the brain to the neuromuscular junction. This initiates an action potential which propagates along the muscle fibre and depolarizes the sarcolemma thus constituting the myoelectrical signal; 2) the excitation-contraction coupling which links the excitation of the sarcolemma to the contractile mechanism; 3) energy metabolism supplying the contractile mechanism. Several studies have shown that the central activation remains sufficient during a fatiguing voluntary muscle contraction (Merton 1954; Bigland-Ritchie and Woods 1984; Bigland-Ritchie et al. 1986b), and that the changes in force generating capacity are primarily associated with peripheral changes beyond the neuromuscular junction.

Several studies have used muscle biopsies to follow the metabolic changes during exercise (Gollnick et al. 1974a,b; Tesch and Karlsson 1977; Sjöholm 1983; Hultman and Spriet 1986). This histochemical technique gives essential information about absolute concentrations of a large number of metabolites, fibre type distribution and enzyme activity. However, muscle biopsy has certain disadvantages in addition to the invasive approach. The number of samples that can be obtained during an experiment is limited and the rather small sample volume may not be representative due to the spatial heterogeneity of the muscle. Errors due to the fast decomposition of certain metabolites before freezing the muscle sample and the fact that the muscle has to be relaxed for some seconds during the biopsy might affect the experimental results of such studies.

In the past twenty years nuclear magnetic resonance spectroscopy (NMR spectroscopy) has been increasingly used to study the energy metabolism of muscle. The non-invasive approach allows measurements of the relative metabolite concentrations during contraction with a relatively high time resolution and a sample volume that is several times larger than that obtained by the biopsy technique. However, the low sensitivity and sample volume localization can be critical problems.

The surface electromyogram (SEMG) shows the myoelectrical signals appearing at the skin surface, as a summation of action potentials generated by the currently active motor units. The SEMG electromyogram is usually recorded by surface electrodes attached to the skin avoiding the discomfort associated with the insertion of needles. The SEMG provides general information about the myoelectrical activity in a muscle, where needle EMG is used to study the activity of a single or a few motor units.

Although exercising muscle has been extensively studied, the physiology of exercising muscle is still not fully understood. Studies of the myoelectrical activity during fatiguing exercise and recovery have produced valuable results, but have often led to a number of theories about the underlying and contemporaneous metabolic changes because the relationship between the myoelectrical activity and the muscle metabolism of exercising muscle has not been thoroughly investigated. Knowledge of the relation between energy metabolism and myoelectrical activity during exercise and recovery could add important new aspects to the understanding of the complex nature of muscle contraction in healthy muscle and in diseased muscle of unknown etiology. The issue can be expressed as the following question: "how are the changes in the myoelectrical activity during exercise associated with the metabolic changes?", leading to the overall aim of this study:

The primary aim of this study was to combine ^{31}P NMR spectroscopy and SEMG to investigate how the myoelectrical changes are related to human skeletal muscle metabolism during exercise and recovery.

^{31}P NUCLEAR MAGNETIC RESONANCE SPECTROSCOPY OF MUSCLE

Primary energy processes of skeletal muscle

The immediate energy needed for muscle contraction and a number of other processes is produced by the breakdown of adenosinetriphosphate (ATP). Regeneration of ATP is achieved by three major energy sources, dependent on work intensity, duration and muscle fibre type:

- Splitting of phosphocreatine (PCr) into ATP and inorganic phosphate (P_i) (the creatine kinase reaction).
- Anaerobic decomposition of glucose and glycogen with lactate formation and a decrease in pH.
- Aerobic decomposition of carbohydrates and free fatty acids.

One of the earliest NMR studies of intact skeletal muscle was obtained from rat hind limbs (Hoult et al. 1974) and later from human skeletal muscle (Ross et al. 1981), and the observable phosphorous metabolites were: P_i, PCr and ATP. A ^{31}P NMR spectrum obtained from a healthy human skeletal muscle is shown in figure 1.

The splitting of the ATP in the ^{31}P NMR spectrum shown in figure 1 into two so-called doublets (α,γ) and one triplet (β) is a result of interaction between adjacent spinning nuclei, and is called spin-spin coupling or j-coupling. Just above the noise level it is sometimes possible to observe the resonances for phosphomonoesters (PME) and phosphodiesters (PDE) as indicated in figure 1.

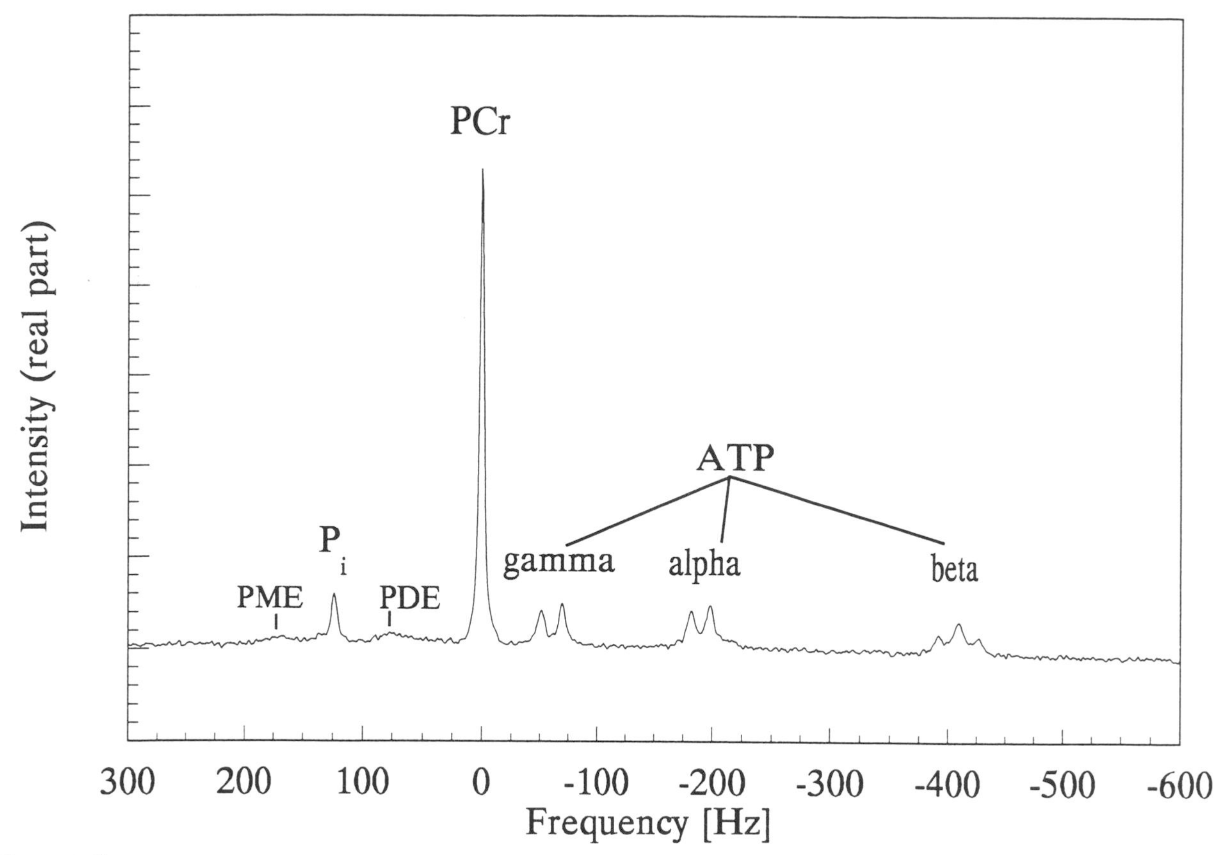

Figure 1. ^{31}P NMR spectrum of the normal human anterior tibial muscle at rest, obtained with a 40x80 mm surface coil and an adiabatic sech/tanh radio frequency pulse with 64 acquisitions.

Measurement of pH is possible with ^{31}P NMR spectroscopy because the chemical shift of P_i relative to the PCr resonance is dependent on the intracellular proton concentration (Moon and Richards 1973) and can be calculated empirically. The PCr is used as an internal chemical shift reference because the pK of PCr is about 4.6 (Gadian 1982) well below the physiological pH range. The integrated resonances in the ^{31}P NMR spectrum are proportional to the molar concentration of the metabolites.

The relative sensitivity of the ^{31}P NMR experiment is about fifteen times less than the ^{1}H NMR experiment in spite of the similar natural abundance of the isotopes in living muscle tissue (Gadian 1982). Furthermore, the PCr concentration in human skeletal muscle is in the magnitude of 30 mM (millimoles x l^{-1}) while the proton concentration is 110 M. This may cause problems in determination of the phosphorous metabolite intensities due to the low signal-to-noise ratio of the signals. The low sensitivity will consequently limit the time resolution because a number of spectra usually have to be averaged to reduce noise contribution.

Insufficient sensitivity of the ^{31}P NMR experiment limits the time resolution and hinders the estimation of the P_i-resonance which sometimes diminishes below the noise level (Taylor et al. 1983; Arnold et al. 1984a; Miller et al. 1988; Newham and Cady 1990; II) during recovery. Another general problem regarding ^{31}P NMR spectroscopy exercise studies is the sample volume localization of the examined muscle, which may cause serious problems due to undesired contribution from other muscle groups. Manual evaluation of the ^{31}P NMR spectra is usually an extremely time-consuming process involving several processing steps. Spectroscopy studies need to have a more automated and operator independent method to process the large number of spectra obtained in exercise studies.

The use of ^{31}P NMR spectroscopy in studies of skeletal muscle has been excellently reviewed in the literature (Dawson et al. 1977; Bárány and Glonek 1982; Taylor et al. 1983; Chance et al. 1985; Burt et al. 1986; Chance et al. 1986; Sapega et al. 1987). The resting levels of PCr, P_i and pH are in general invariant in normal human skeletal muscle, and the changes in PCr intensity during exercise and recovery are the reverse of the changes in P_i intensity. Furthermore, the resting PCr/P_i ratio has been shown to be correlated to the muscle fibre type composition, showing a higher PCr/P_i resting level the larger the amount of fast twitch fibres within the muscle (Meyer et al. 1985; McCully et al. 1989; Kushmerick et al. 1992; Bernus et al. 1993). The PCr/P_i ratio has been shown to mirror the steady-state oxidative potential of the muscle, and is correlated to oxygen delivery and the ADP concentration (adenosinediphosphate) which is thought to control the creatine-kinase reaction (Chance et al. 1981). Changes in ATP intensity are usually not observed even during exhausting exercise, though decreased ATP levels have been reported in finger flexor muscle, which were associated with extreme PCr losses (Taylor et al. 1983). Splitting of the P_i resonance into two separate peaks seems to be attributable to slow- and fast-twitch fibres having different pH (Park et al. 1987; Vandenborne et al. 1991; Mizuno et al. 1994).

The PCr recovery after exercise occurs in an exponential manner and the recovery rate is related to the oxidative capacity (Taylor et al. 1983; Arnold et al. 1984a). Severe acidosis prolongs the recovery (Arnold et al. 1984a). The PCr and pH recovery are prevented by blood flow occlusion (Thomsen 1988) and both remain constant thus indicating that glycolytic activity stops immediately at the end of muscle contraction. Recovery is initiated by reperfusion and indicates that recovery of PCr primarily depends on oxidative metabolism (Harris et al. 1976).

SURFACE ELECTROMYOGRAPHY

Quantitative surface electromyography

Quantitative analysis of the SEMG uses signal parameters describing the amplitude- and frequency-developments of the SEMG during exercise which are usually estimated by computerized algorithms using segments of the time-discrete sampled SEMG. The root-mean-square (RMS) is generally used as an amplitude parameter. Alternatively the rectified and integrated EMG (IEMG) is used. The median power frequency (MPF) and the mean frequency have mainly been used to describe changes of the SEMG power spectrum during exercise. A review of SEMG is given in (de Luca 1984)

The SEMG changes dynamically during exercise and recovery (Broman et al. 1985). The RMS of the SEMG increases during fatiguing exercise of submaximal intensity (Häkkinen and Komi 1983) which is mainly related to motor unit recruitment (Bigland-Ritchie et al. 1986a). During fatiguing exercise a frequency shift of the SEMG power spectrum towards lower frequencies has been observed, and is mainly related to the shape of the motor unit action potential and to a lesser extent to the motor unit firing rate (Lindström and Magnusson 1977). In a number of studies the frequency shift of the SEMG has been correlated to the muscle fibre conduction velocity, showing a highly linear relationship (Arendt-Nielsen and Mills 1985; Eberstein and Beattie 1985; Zwarts et al. 1987). A decreasing conduction velocity is thought to be due to lactate and/or proton accumulation (Lindström et al. 1970; Mortimer et al. 1970). Induced ischaemia has been shown to prevent recovery of MPF after exercise (Hara 1980; Mills 1982). In addition it has been shown that lactate concentration correlated linearly to the mean power frequency of the EMG (Tesch and Karlsson 1977).

A previous study combined ^{31}P NMR spectroscopy and SEMG (Miller et al. 1987), but the integrated EMG was only measured during recovery from fatiguing exercise and myoelectric spectral parameters were not examined.

APPROACH OF THE PRESENT STUDY

To address the question of how the myoelectrical signals are associated with the metabolic changes during exercise and recovery using simultaneous ^{31}P NMR spectroscopy and SEMG, the study was divided into i) a methodological part concerning technical aspects of simultaneous ^{31}P NMR spectroscopy and SEMG and ii) an experimental part applying the technique in vivo:

i) Redesign of a non-magnetic calf muscle ergometer (Quistorff et al. 1990) for use in a whole-body NMR magnet (I,II). Sensitivity enhancement of ^{31}P NMR spectroscopy by surface coil and radio frequency pulse optimization (II) and determination of sample volume localization performed by ^{31}P and ^{1}H NMR imaging techniques using the developed surface coil and radio frequency pulse (II). Design of SEMG in strong magnetic fields and subsequent examination of the interactions with the ^{31}P NMR spectroscopy (I,II). Application of signal-processing of ^{31}P NMR spectroscopy and SEMG data (I,II).

ii) The relation of ^{31}P NMR spectroscopy and SEMG parameters in healthy volunteers was examined during static exercise and recovery at various work intensities (I,III). Recovery SEMG was acquired from brief contractions during recovery. The work intensity ranged from light non-exhausting exercise to strenuous exhausting exercise with considerable anaerobic activity so that the SEMG response to the different energy processes of exercise was examined.

A patient with the rare congenital metabolic error, McArdle's syndrome, was included in the study (III). The patient's inability to produce lactate and accumulate protons due to a defect in muscle glycogen breakdown (McArdle 1951), makes it possible to examine the SEMG response during exercise and recovery without the influence of lactate and proton accumulation which are thought to be important contributors to myoelectrical changes. Examination of the patient with McArdle's syndrome also gives a good opportunity to investigate causal relationships between the SEMG changes and proton and lactate accumulation.

The investigation included patients with fibromyalgia to see if the method could be useful for diseases of unknown etiology. Fibromyalgia patients have complaints about muscular fatigue, generalised muscular pain and have reduced muscular force and low physical activity levels (Jacobsen and Danneskiold-Samsøe 1987; Bengtsson and Henriksson 1989; Clark et al. 1992a), but it was unknown if the reduced muscular force was due to a reduced muscle size. Classification of fibromyalgia has been reviewed in (Yunus et al. 1981; Wolfe et al. 1990) and implicate tenderness in a number of certain anatomical sites. Muscle biopsy studies report no abnormalities but a higher occurrence of unspecific degenerative changes (Yunus 1989; Schrøder et al. 1993).

Despite a large number of studies, including SEMG studies (McBroom et al. 1988; Zidar et al. 1990; Elert et al. 1992) and ^{31}P NMR studies (Blecourt et al. 1991, Jacobsen et al. 1992), it has not been possible to detect any major abnormalities. However, a decreased content of energy-rich phosphates (Bengtsson et al. 1986) and inhomogeneous oxygen saturation (Lund et al. 1986) have been reported in fibromyalgia, but have not been additionally confirmed. It is uncertain whether the disease is related to muscle metabolism, the excitation contraction coupling, neurogenic factors or other factors. Most studies have been performed on resting muscle or during mild exercise, but in agreement with (Kushmerick 1989) a dynamic stress test including exercise and recovery is necessary to differentiate normal from abnormal. Less extreme exhaustion levels of PCr and pH have though been reported from a study using a work test (Jacobsen et al. 1992). Recently SEMG and ^{31}P NMR spectroscopy were combined in the evaluation of fibromyalgia (Roy 1994). No differences in SEMG or NMR parameters during exercise and recovery were found between sedentary controls and patients, but an analysis of the correlation between NMR and SEMG parameters was not performed.

A question to be addressed could be: do the fibromyalgia patients have normal motor unit activity in relation to muscle metabolism during exhausting exercise and recovery? Further, do fibromyalgia patients have reduced muscular force per muscle area compared with healthy subjects and has the low physical activity some significance?

To address these questions, patients with fibromyalgia and matched controls were examined using simultaneous ^{31}P NMR spectroscopy and SEMG during exhausting exercise and recovery. The maximum muscular force and maximum transverse muscle area were estimated using ^{1}H NMR imaging (IV, V). The controls were matched to the patients as regards age, gender and daily physical activity levels (the controls are termed "sedentary controls").

2. METHODOLOGICAL STUDIES

In a correlation study of the myoelectrical activity and metabolism of exercising and recovering skeletal muscle the measurements must be of sufficiently high quality and it must be assured that the processes observed originate from the same muscle. The technical aim was to ensure reliable measurements of SEMG and ^{31}P NMR spectroscopy originating solely from the skeletal muscle concerned.

MUSCLE MODEL AND ERGOMETER

The anterior tibial muscle seemed to be a very suitable muscle model when performing simultaneous ^{31}P NMR spectroscopy and SEMG during exercise, because the muscle is superficial and easily accessible and the primary agonist of static ankle dorsiflexion. Ankle dorsiflexion was easily performed inside the narrow bore of a whole-body magnet used for ^{31}P NMR spectroscopy in this study, and both static and dynamic exercise are possible with the ergometer following a slight modification (I). All exercise protocols used low level isometric contractions with constant angle and force because the maximum torque about the ankle joint changes with the angle and alterations in force affect the myoelectrical parameters (Broman et al. 1985). Dynamic exercise would increase the possibility of interaction between the SEMG and ^{31}P NMR spectroscopy in spite of the technical design of SEMG during simultaneous ^{31}P NMR spectroscopy (I). The ergometer system provided the subject with visual feedback of the pedal angle and was strictly supervised during the whole experiment so that force variation was negligible. The pedal axis could be adjusted to align with the ankle joint axis, so that measurements of the maximum voluntary contraction force (MVC) of the anterior tibial muscle could be performed. The MVC was estimated from three trials as the maximum torque the subject was capable of resisting for 3 seconds with a 10° plantar flexion. The ergometer principle is illustrated in figure 2.

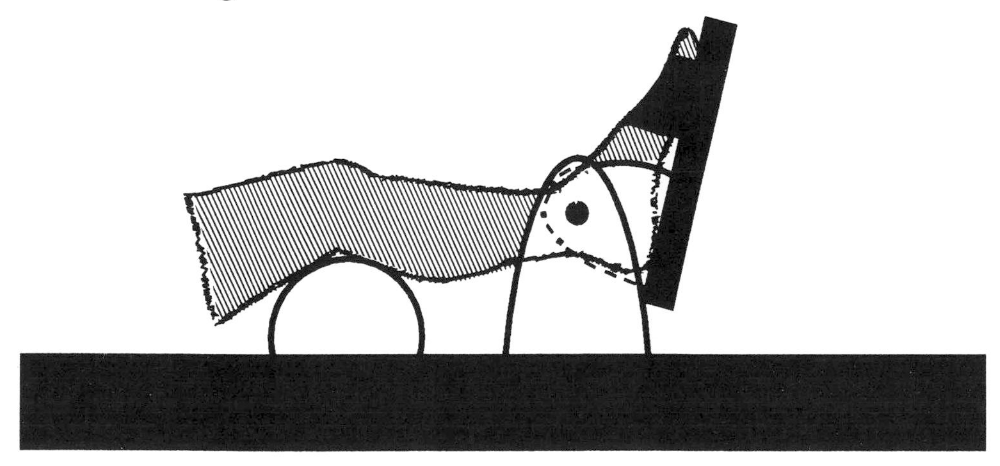

Figure 2. A diagram of a pedal ergometer suitable for isometric contractions of the anterior tibial muscle.

MUSCLE MORPHOLOGY AND ACTIVATION USING NMR IMAGING

Using ^{1}H NMR imaging, the morphology of the anterior tibial muscle was investigated, which was essential for the development of a suitable surface coil for the ^{31}P NMR spectroscopy. The selective activation of the anterior tibial muscle during dorsiflexion in the ergometer was verified using ^{1}H NMR imaging, where transaxial images obtained before and after exhaustive ankle dorsiflexions showed a significant signal increase alone in the anterior muscle (II) as shown in figure 3. The increased image intensity in the anterior tibial muscle produced by an increased T_2-relaxation time is possibly due to the redistribution of water within the exercising muscle (Fleckenstein et al. 1988, Fisher et al. 1990). Transaxial images were obtained from all subjects to determine the maximum transverse area of the anterior tibial muscle.

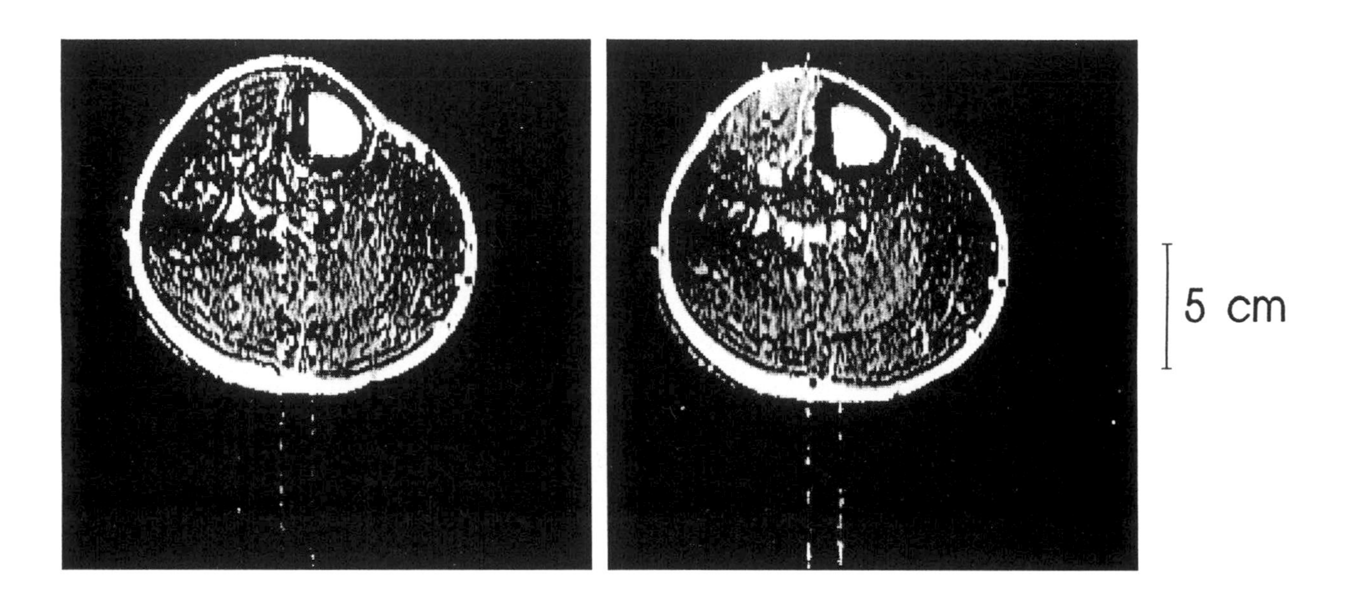

Figure 3. Transaxial spin echo images of the lower limb of a healthy volunteer (TE=90 ms; Tr=2000 ms; matrix = 256x256; slice thickness=5mm; 170 mm Helmholtz coil) before (left) and immediately after (right) exhausting ankle dorsiflexions. The T_2 relaxation time of the anterior tibial muscle prior to exercise (29.6 ms $\pm$ 1.6 ms) was significantly less than after exercise (36.6 $\pm$ 1.6 ms) $P<0.001$ (Student's two-tailed t-test).

^{31}P NMR SPECTROSCOPY: SENSITIVITY ENHANCEMENT AND VOLUME LOCALIZATION

Surface coil

Examination of a superficial skeletal muscle by ^{31}P NMR spectroscopy using a surface coil was considered the optimum technique, because this method, in contrast to volume selective spectroscopy (Buchtal et al. 1989; Luyten et al. 1989), does not use magnetic field gradients that could disturb the SEMG. In addition, the volume selective methods have lower sensitivity and are motion sensitive.

A one-turn 40x80 mm elliptical surface coil was constructed according to the guide lines obtained by ^1H NMR imaging and this also facilitated attachment of SEMG surface electrodes inside the loop of the coil. The special design minimized losses and was specifically insensitive to movements and cable contact. A gain of 45% in signal-to-noise ratio was obtained compared to standard techniques which utilized a 40 mm circular surface coil (II). The result corresponds to the expected improvement when doubling the sensitive volume, but indicates that surface coil design is crucial in ^{31}P NMR spectroscopy of skeletal muscle concerning sensitivity.

Radio frequency pulse and sample volume

Generation of the oscillating B_1 magnetic field by a radio frequency pulse with the resonance frequency when using a surface coil has the disadvantage of field inhomogeneity. A square-modulated radio frequency pulse is generally used but this does not optimally excite all the spins in the sensitive volume. To achieve optimum excitation of the muscle tissue in the sensitive volume of the surface coil two different types of adiabatic radio frequency pulses were examined (II). Using an adiabatic radio frequency pulse, it was possible to excite equally all the spins of a sample. The adiabatic sech/tanh radio frequency pulse (amplitude and phase modulated) was the most suitable and improved the signal-to-noise ratio by 56% compared to a square-modulated radio frequency pulse. The sech/tanh pulse was also superior to the other adiabatic sin/cos pulse and the square-modulated radio frequency pulse as regards linearity of the spectral excitation of a skeletal muscle ^{31}P NMR spectrum (II). The spatial selection of the surface coil and the sech/tanh pulse were examined by ^{31}P NMR imaging, and were found to cover the adult anterior tibial muscle with a half-ellipsoid formed volume with the axis 40 X 80 mm (the geometry of the surface coil) and 20 mm deep (II). T_2 relaxation effects during the radio frequency pulse transmittance were negligible because the pulse length (0.640 ms) was small compared to the shortest T_2 relaxation time of the examined metabolites (8 ms of ATP-beta (Thomsen et al. 1989a)).

The power requirements for the sech/tanh adiabatic pulse were much higher than the square-modulated pulse but the calculated mean power dissipation of the sech/tanh pulse was considerably smaller than the maximum recommended values in the literature (INPA/INIRC 1991) and heat absorption in the sensitive volume of the coil was therefore considered negligible. Optimization of the transmitter power can be done in advance in contrast to the square-modulated pulse where optimization of the transmitter power must be performed for every experiment to obtain maximum signal-to-noise ratio. The high power requirements using the implemented adiabatic pulse (II) may exceed the power available in some NMR systems. A longer pulse length would considerably reduce the power requirements, but the T_2 relaxation times are very short, especially of ATP (Thomsen et al. 1989a), and may lead to a substantial signal loss in the ATP-region. The sensitive sample volume of the coil in combination with the adiabatic pulse matched the physical dimensions of the average-sized anterior tibial muscle of adults well, but the physical dimensions of the coil must be reconsidered in clinical applications, e.g. severe muscle atrophy, to avoid signal contribution from other muscle groups. Muscle atrophy was not present in the group of patients with fibromyalgia and the patient with McArdle's disease and the same coil dimensions could be used.

The optimum interval between acquisition of two consecutive ^{31}P NMR spectra using the adiabatic pulse, if the focus is on signal-to-noise ratio, was seven seconds and exceeds the T_1 relaxation time of the human skeletal muscle metabolites at the specific B_0 field strength (Thomsen et al. 1989b). A compromise was made between the signal-noise ratio of the spectra and time resolution, and ^{31}P NMR spectra obtained during exercise were an average of two consecutive spectra. During muscular rest, time resolution is not critical and a high quality spectrum was obtained with an average of 64 acquisitions.

^{31}P NMR postprocessing

Manual processing of the ^{31}P NMR spectra involves several steps: The FID is multiplied with a decaying exponential or Gaussian function to reduce noise contribution. Frequency analysis is performed by FFT (fast Fourier transformation) and the spectrum must be phase corrected and the intensities integrated to obtain the relative metabolite concentrations. In some cases an additional baseline correction must be applied before peak integration, to remove the broad resonance (hump) underlying essential parts of the spectrum. The broad resonance is due to a disturbed initial part of the FID, probably due to very fast decaying metabolites (Gadian 1982).

In addition to specific time- and frequency-domain algorithms (Hoult et al. 1983; Nelson and Brown 1987) two major types of promising time-domain postprocessing techniques have been proposed: non-linear least-square approximation using a model of the signal (Van der Veen et al. 1988) and polynomial approximation using linear prediction (Diop et al. 1992). Polynomial approximation needs less operator involvement, but the non-linear least-square techniques have the possibility of applying prior knowledge of the signals and discarding the initial part of the FID in the analysis to avoid humps in the frequency-domain (Van der Veen et al. 1988, de Beer and van Ormondt 1992). This is especially important when dealing with signals of low signal-to-noise ratio (de Beer and van Ormondt 1992). Studies of in vivo NMR signal processing techniques preferred the non-linear least-square technique to linear prediction methods (Joliot et al. 1991, Ho et al. 1992, de Beer and van Ormondt 1992).

The non-linear least-square fitting procedure VARPRO (Van der Veen et al. 1988) was considered the most suitable for processing in vivo ^{31}P NMR spectra of skeletal muscle. Prior knowledge of phases, linewidths and amplitudes was utilized, and especially the constant chemical shift and amplitude ratios of the ATP doublets and triplets could be constrained or linked to each other due to the known j-coupling (II). The initial data points of the FID were omitted and the broad resonance was efficiently rejected. The high quality ^{31}P NMR spectrum obtained during rest was used to estimate important general prior knowledge parameters, utilized in processing of the exercise and recovery spectra with much lower signal-to-noise ratio. The spectral estimates using the VARPRO fitting procedure had relatively low errors (II) and all metabolites could be estimated during rest, exercise and the entire recovery period. The quality of the spectral processing was controlled by inspecting the experimental, estimated and residual spectrum which are shown in figure 4.

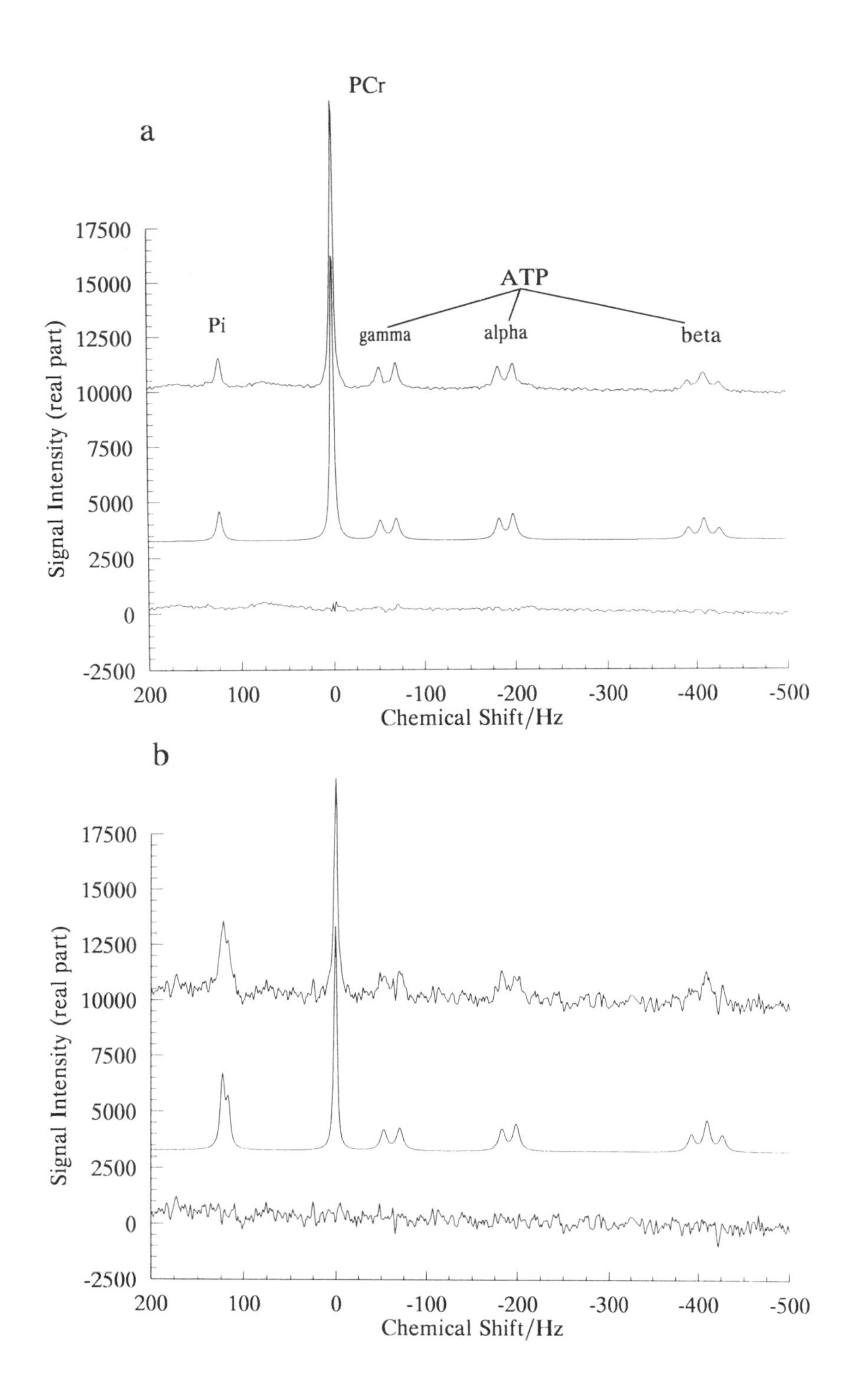

Figure 4. VARPRO fitting performance. Panel a: 64 acquisition ^{31}P NMR spectrum of the human anterior tibial muscle obtained during rest. Panel b: 2 acquisition spectrum obtained during exercise. Top (a,b):(Fast Fourier transform of the unprocessed FID). Middle(a,b): VARPRO estimate with 9 (panel a) and 10 resonances (panel b). From left: P_i (splitting P_i during exercise), PCr, ATP(γ), ATP(α) and ATP(ß). Bottom (a,b): Residual signal.

Computerized algorithms were constructed to calculate metabolite intensities and pH. The chemical shift of the splitting P_i resonance necessary to calculate pH was calculated using the weighted mean of the two resonances. The PCr and P_i intensities were normalized to the sum: PCr+P_i, avoiding possible errors due to decreasing sensitivity during an experiment. pH was calculated empirically from the chemical shift of P_i to the PCr resonance acting as a chemical shift reference according to the formula used in (Moon and Richards 1973). The pH error of estimation due to digital resolution was less than 0.01 pH units. The recovery of PCr and pH was fitted to a mono-exponential function (III).

No significant ATP changes from the resting level could be detected during exhausting exercise and recovery in any of the healthy volunteers, patients with fibromyalgia or the McArdle patient, and the parameter was consequently withdrawn from the further analysis (II,III,IV). Major ATP changes would have been detected in spite of the fact that ATP estimation errors were somewhat larger than the PCr and P_i estimation errors (III), and only minor ATP changes could have occurred.

Preliminary in vivo experiments (II) demonstrated that the recovery of PCr and P_i were clearly affected by the brief test contractions necessary to obtain recovery SEMG, as judged by the intensity fluctuation during periods with and without test contractions. The fluctuations of PCr and P_i and consequently the PCr/P_i ratio were larger the higher the contraction level, whereas the pH recovery seemed less affected (III). An attempt to obtain ^{31}P NMR spectra which did not coincide with test contraction periods by triggering the acquisition was unsuccessful due to instrumental and software limitations.

The compromise between quality and time resolution in ^{31}P NMR spectroscopy will always exist. The time resolution of the ^{31}P NMR spectroscopy during exercise was approximately 20 s where the net measurement time was 14 sec (2 acquisitions each of seven sec) and the rather large residue was used for data storage and could not be modified in the NMR system used (I,II). The averaging of two consecutive spectra, combined with phase-cycling to compensate for offset-errors in the receiving system, was chosen to obtain sufficiently accurate estimates of the metabolites concerned. The attained time resolution is judged to cover the main features of the dynamic metabolic changes in similar exercise protocols. However, when the metabolic changes during a contraction cycle (Quisttorf et al. 1988) or high intensity exercise are investigated, the time resolution and consequently the sensitivity must be improved considerably. Further optimization of the NMR sensitivity is possible by utilization of proton decoupling (Ernst 1966).

SURFACE ELECTROMYOGRAPHY DURING ^{31}P NMR SPECTROSCOPY

The use of electromyographic recordings during NMR spectroscopy raises several important aspects that must be addressed: the effect of surface electrodes on the magnetic field homogeneity, noise contribution to the ^{31}P NMR spectroscopy via the SEMG leads, and the effect of ^{31}P NMR spectroscopy on the SEMG.

The technical solution involved the use of bipolar non-magnetic surface electrodes, shielded leads and radio-frequency filters as described in (I,II). The surface electrodes did not impair the magnetic field homogeneity and no effects on the sensitivity of the ^{31}P NMR spectroscopy could be detected. When the surface electrodes were placed beside the surface coil on the longitudinal axis of the anterior tibial muscle, the NMR spectroscopy did not affect the SEMG (I), but a short transient generated by the radio frequency pulse appeared when the electrodes were placed inside the surface coil (II). Nevertheless, all experiments (II,III,IV) were performed with the surface electrodes placed inside the surface coil to ensure as far as possible that the myoelectrical and metabolic processes originated from the same muscle volume.

The position of the SEMG electrodes in relation to the innervation zone on the anterior tibial muscle was judged to have minor effect on the estimation of the relative spectral SEMG changes during exercise as previously shown (Roy et al. 1986). Roy et al. (1986) showed that the highest values of MPF always occurred near the innervation zone, and decreased proportional to the distance from the innervation zone, but the rate of MPF changes during static exercise was not effected by the electrode position.

A data acquisition computer outside the radio frequency shielded cabin surrounding the magnet was used to continuously sample and store the SEMG. The digital sampling and filtration were performed in accordance with the frequency content of the SEMG (II). A trigger signal describing the precise time of the ^{31}P NMR spectroscopy was concurrently sampled because an accurate time base of the ^{31}P NMR spectroscopy was essential when correlating with the SEMG.

SEMG postprocessing

Segmentation of the stored SEMG was performed every five s. Signals containing transients from radio frequency were easily avoided, because the simultaneously sampled NMR time base supplied the precise time of every radio frequency pulse. The segment length was chosen according to stationarity of SEMG during static exercise (Inbar and Noujaim 1984; II) and to obtain accurate signal parameter estimates.

The SEMG power spectrum was estimated by autoregressive modelling (AR) using the autocorrelation method, proved to be efficient in spectral analysis of SEMG (Inbar and Noujaim 1984, Paiss and Inbar 1987). The AR model order was sufficient to estimate the characteristic spectral envelope when compared to a spectrum obtained by FFT (II). The MPF was selected because it has been shown that the MPF was less sensitive to noise contributions than the mean power frequency (Stulen and de Luca 1981). The root-mean-square (RMS) of the SEMG segments could be directly estimated using the zero time-lag autocorrelation calculated in the AR-model to estimate the SEMG power spectrum. Estimation of myoelectric amplitude and frequency parameters such as RMS and MPF using AR-modelling is considered an effective way to estimate basic parameters in quantitative SEMG. The selection of segment length, sample frequency, filtration and model order is important (I,II).

The technical part of the present study (I,II) proved that quantitative SEMG during ^{31}P NMR spectroscopy in a whole-body magnet surrounded by a radio frequency shielding cabin was possible without deterioration of the ^{31}P NMR spectroscopy, and placing SEMG electrodes inside the surface coil ensured as far as possible that the SEMG as well as the ^{31}P NMR spectroscopy originated from the anterior tibial muscle. The ^{31}P NMR spectroscopy sample volume was approximately 33 ml large and was judged to constitute a representative volume of the normal adult anterior tibial muscle. The activation of the anterior tibial muscle during ankle dorsiflexions was verified by proton imaging and indicated that the SEMG during isometric ankle dorsiflexion originated only from the anterior tibial muscle.

The technical solutions doubled the sensitivity of the ^{31}P NMR spectroscopy compared to initial conditions, and indicate that a dedicated surface coil design and the choice of radio frequency pulse are crucial when performing ^{31}P NMR spectroscopy of skeletal muscle during exercise and recovery. Further, ^{31}P NMR spectroscopy studies of exercising and recovering muscle with high time resolution seems unrealistic without at least a semi-automated post-processing method such as the VARPRO fitting routine that was used in this study.

3. IN VIVO STUDIES

The purpose of the in vivo studies (I,III,IV) was to obtain experimental data about myoelectrical and metabolic variables during rest, exercise and recovery. The metabolic parameters that were estimated were PCr, P_i, PCr/P_i and pH. ATP were not included in the analysis due to a complete lack of change in any of the examined healthy volunteers or patients (II,III,IV). The myoelectrical parameters were RMS and MPF that describe amplitude changes and SEMG spectral changes, respectively. The results were used in a correlation analysis to examine the myoelectrical - metabolic relationships at different work intensities in healthy volunteers (III), the complete lack of proton and lactate accumulation in the patient with McArdle's syndrome (III), and to study the myoelectrical-metabolic relationship and the maximum muscle force per muscle area in the fibromyalgia patients and the matched sedentary controls (IV,V).

MATERIAL AND EXERCISE PROTOCOLS

Young, age-matched and energetic male volunteers were selected to minimize a possible influence of sex and age (III). The healthy volunteers had no history of diseases related to metabolism or the circulatory system, and did not receive any kind of drugs. All the volunteers participated in sport regularly. One of the volunteers performed all the exercise protocols three times to obtain an estimate of the experimental variance of measurement.

A young male patient with McArdle's syndrome was examined once. No lactate was detected when the patient did an ischaemic arm exercise test. A muscle-biopsy study confirmed the finding that the enzyme myophosphorylase a and b (glycogen phosphorylase) was absent.

A group of female patients with fibromyalgia and a group of age-matched sedentary female controls were examined (IV). The controls were selected to optimally match the daily energy consumption levels of the patients (IV). The sedentary controls did not receive any medication and had no history of diseases affecting the cardiovascular or muscular system.

All experiments used submaximal voluntary isometric contraction of the anterior tibial muscle. In the first study with healthy volunteers (I) a constant submaximal torque was applied and recovery was not examined. In the subsequent in vivo study three isometric levels were applied in the experiments with healthy volunteers (III) : 5%, 10% and 30% of the MVC and are termed MVC-5, MVC-10 and MVC-30, respectively. An isometric contraction with constant angle and torque was sustained until exhaustion and brief intermittent contractions were performed in the 30 min recovery period to obtain recovery SEMG (III). The MVC-5 was aborted if the volunteers were able to perform the exercise for longer than 30 min. Exhaustion was defined as the time at which the subject was no longer capable of holding a constant pedal angle, and the pedal was dropped. The McArdle patient and the fibromyalgia patients performed only the MVC-10 protocol.

The studies were conducted in accordance with the Helsinki Declaration II. Informed consent was obtained from all subjects, and the experimental protocols were approved by the local ethical committee.

RELATION OF METABOLISM AND MYOELECTRICAL ACTIVITY IN YOUNG HEALTHY VOLUNTEERS AND A PATIENT WITH McARDLE'S SYNDROME

Exercise endurance and maximum voluntary contraction force

The MVC and maximum cross-sectional muscle area were similar in the examined heathy volunteers (III). The endurance time was longer the lower the contraction level, as expected, but large variations of endurance time were seen between subjects (III). The MVC of the McArdle patient corresponded to the healthy volunteers (III), but the endurance time was considerably shorter than the average healthy volunteer performing MVC-10.

The relation between endurance and isometric contraction levels in a number of arm and leg muscles has been reviewed by (Rohmert 1962). It was shown that the endurance time increased in a hyperbolic manner as the isometric contraction force was reduced. The formula is applicable down to 15% MVC and endurance time less than 10 min, and seemed to be valid for all the examined muscle groups (Rohmert 1962). The endurance time of 15% MVC static exercise was very long, and Rohmert (1962) stated that the endurance time was practically infinite for a contraction force of less than 15% MVC. The endurance time of MVC-30 of the anterior tibial muscle in healthy volunteers was on average 4.4 min (III), compared to 2.5 min calculated from the empiric formula in (Rohmert 1962). The theoretical infinite endurance time of a 10% MVC static contraction as stated by Rohmert (1962) could not be supported by the findings of the present study, because the endurance time of MVC-10 in the healthy volunteers was on average 13.2 min (III). On the other hand, it seemed that a 5% MVC level of the anterior tibial muscle had a practically infinite endurance time, because most volunteers could sustain a 5% MVC contraction for more than 30 min (III). Many factors may contribute to the variation in endurance time, including the method of MVC estimation, fibre type composition of the muscle and physical condition.

The experimental variance of the SEMG and NMR measurements

An estimate of the precision of the NMR and SEMG measurements was examined using the intra-subject and inter-subject variability (III). NMR and SEMG parameters obtained from the same subject on different days did not show a statistically significant difference from those obtained from the other healthy volunteers as regards mean and variance. In addition low coefficients of variation were observed in identical experiments with the same volunteer.

The metabolic resting levels are constant and the initial myoelectrical parameters are dependent on the applied contraction level

All SEMG and ^{31}P NMR spectroscopy parameters during rest, exercise and recovery are shown for the applied contraction levels in figure 5. The initial levels of RMS and MPF were higher the higher the contraction level applied, as earlier noticed (Broman et al. 1985). Broman et al. (1985) also showed that the conduction velocity was proportional to the contraction level. It was suggested that the proportionality of RMS was due to the recruitment of systematically larger motor units and that the proportionality of the conduction velocity was due to recruitment of motor units with larger fibres that reduce the MPF. The resting levels of PCr, P_i and especially pH were invariant and in accordance with earlier NMR studies (Radda 1984, Taylor et al. 1983, Molé et al. 1985) and the pH level was in accordance with biopsy studies (Hultman et al. 1985). The pH resting level of the McArdle patient was normal but the PCr/P_i resting level was nearly 40% higher than the average healthy volunteer (III), which is in contrast to earlier studies where the resting PCr/P_i ratio was similar to that of healthy volunteers (Pryor et al. 1990; Bendahan et al. 1992).

Three different myoelectrical and metabolic responses to static exercise: 5% MVC produced steady-state levels; 10% MVC produced slow changes towards exhaustion with temporary steady-state periods and 30% MVC produced rapid changes towards exhaustion

In general there were three different responses to the applied isometric contraction levels : 5%, 10%, and 30% MVC as illustrated in figure 5. A 5% MVC contraction could be sustained without great effort for 30 min by most volunteers, but some were exhausted in the late part of the exercise. In the non-exhausted cases the metabolic and myoelectric parameters rapidly reached a steady-state level, which was maintained throughout the exercise. A 10% MVC contraction induced marked changes in both metabolic and myoelectrical parameters progressively changing towards exhaustion with a rapid initial decrease of PCr and PCr/P_i, but with a temporary steady-state period where both the myoelectrical and metabolic parameters remained constant. A similar pattern was seen in the subjects exhausted during a 5% MVC contraction. A 30% MVC induced progressive changes in both myoelectrical and metabolic parameters towards exhaustion, and no steady-state periods were observed.

Steady-state periods (or working points) of PCr/P_i observed in the MVC-5 and MVC-10 have also been reported previously (Idström et al. 1985), where oxygen delivery and consumption were shown to be linearly correlated to the PCr/P_i ratio also shown to mirror the current steady-state work intensity. The steady-state working points may therefore indicate periods of exercise where the ATP hydrolysis rate equals the synthesis rate and motor unit recruitment remains unchanged judging by the constant RMS and MPF. Unlike the temporary steady-state periods in MVC-10 and the exhausted subjects in MVC-5, the steady-state periods persisted throughout the 30 min exercise in the non-exhausted subjects in MVC-5 (III). Other studies support the view that an isometric 5% MVC contraction level is a long term steady-state level.

Figure 5. Experimental results of a 5%, 10% and 30% MVC isometric contraction (contraction scheme incorporated in the figures) in normal volunteers. Vertical dotted line indicates time of exhaustion. All NMR data were estimated by the VARPRO fitting routine. The MPF was estimated by a 20th order AR model. During recovery a single 12 sec contraction was performed every minute for 20 minutes, followed by a 10 min period of rest with no contractions.

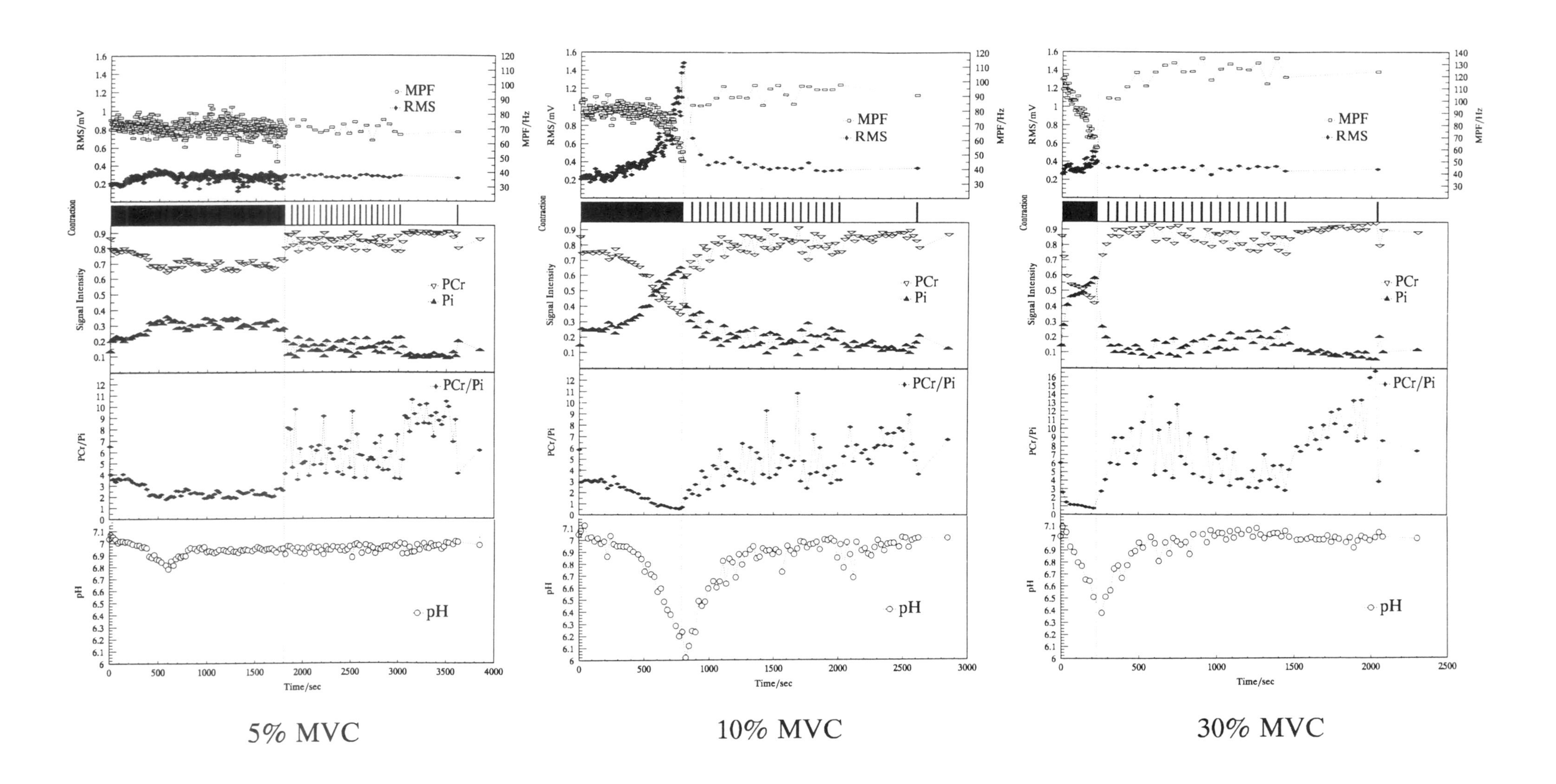

During a 5% MVC static knee extension, blood flow and arterial plasma potassium increased initially to a steady-state level and remained at this level throughout 1 hour of exercise, but during 15% MVC and higher static contraction levels, the blood flow did not increase correspondingly (Saltin et al. 1981, Sjøgaard et al. 1986).

The initial fast decrease in PCr/P_i to a steady-state level in the MVC-5 and to the temporary steady-state level of MVC-10 indicates that some adaptation to the increased energy demand and oxygen consumption is probably regulated by an increased blood flow. Hence, the initial part of the contraction seems to be supplied with energy primarily from the breakdown of PCr because the PCr initially drops very rapidly. Similar steady-state pH levels were seen following a small initial decrease and indicate that some anaerobic glycolytic activity is present at the beginning of the contraction and during the steady-state periods.

The MPF decreased while the pH increased during exercise in the patient with McArdle's syndrome

In the McArdle patient the RMS increased normally during the exhaustive 10% MVC contraction, but the MPF decreased less than in healthy volunteers (the McArdle patient is the same age as the healthy volunteers) compared to the marked changes in RMS, PCr and P_i which were similar to that of the healthy volunteers (III). As expected the pH did not decline in the McArdle patient, but even increased slightly during exercise as described earlier (Ross et al. 1981, Jensen et al. 1990; Bendahan et al. 1992). The pH increased gradually, more than 0.1 unit during the exercise. All parameters during exercise and recovery are shown in figure 6.

The RMS increases hyperbolically with decreasing PCr/P_i during submaximal isometric exercise. Maximum observed motor unit recruitment is associated with a consistent exhaustion level of $PCr/P_i = 0.6$

During exercise the RMS-PCr/P_i correlation fitted well to a hyperbolic function shown in figure 7 (the RMS-PCr/P_i correlation diagram during both exercise and recovery is shown in figure 8a). As the PCr/P_i decreased the initial RMS remained unchanged until the PCr/P_i decreased below a working point region of $PCr/P_i=3$ where both parameters remained constant for a period of time. When the PCr/P_i decreased further, the RMS increased and reached maximum at the exhaustion point: $PCr/P_i=0.6$ (I,III). The exhaustion PCr, PCr/P_i and pH levels were similar for all exhaustive protocols, and the low pH of 6.3-6.4 indicated severe acidosis at the time of exhaustion. The RMS-PCr/P_i relation during the MVC-30 were similar to that of the MVC-10 but changed gradually towards exhaustion and did not show visible working points (II). The RMS-PCr/P_i of the non-exhausting MVC-5 remained at the working point until the contraction ended.

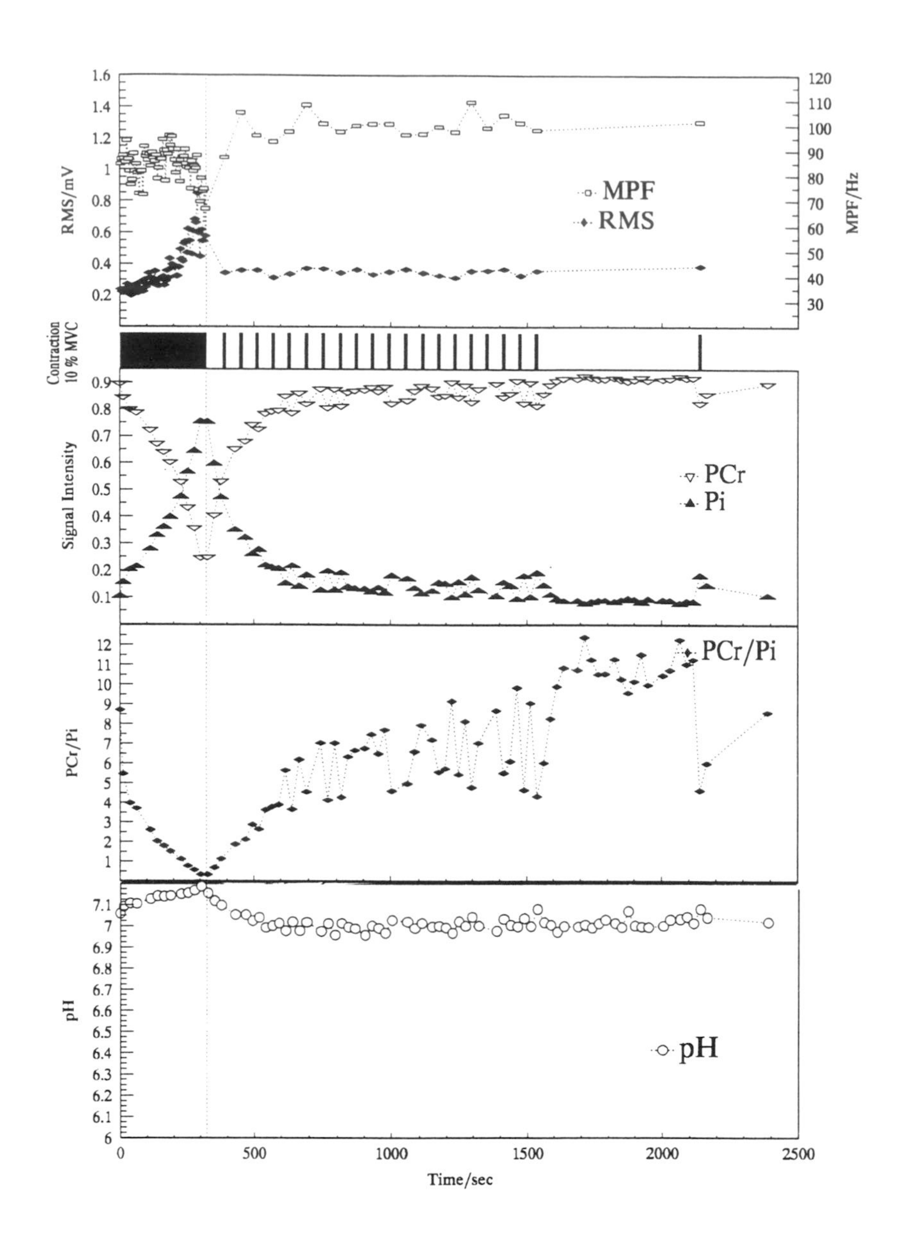

Figure 6. Patient with McArdle's Syndrome. Experimental result of a 10% MVC isometric contraction (contraction scheme incorporated in the figures). Vertical dotted line indicates time of exhaustion. All NMR data were estimated by the VARPRO fitting routine. The MPF was estimated by a 20th order AR model. During recovery a single 12 sec contraction was performed every minute for 20 minutes, followed by a 10 min period of rest with no contractions.

The exhaustion level of $PCr/P_i=0.6$ at which maximum RMS values were observed, which was interpreted as maximum motor unit recruitment associated with maximum reduced MPF (figure 8c), is probably a metabolic limit to a voluntary submaximal isometric contraction. The exhaustion level was observed in all experiments leading to exhaustion in the studies of young healthy volunteers (I,III) and the hyperbolic RMS-PCr/P_i relation and exhaustion level of the McArdle patient were similar to those of healthy volunteers (III). The exhaustion level seems to be valid for a range of isometric contraction levels as it was observed in the MVC-5 (exhausted subjects), MVC-10, MVC-

30 in a test with a constant submaximal torque level performed by a group of healthy young male and female volunteers (I).

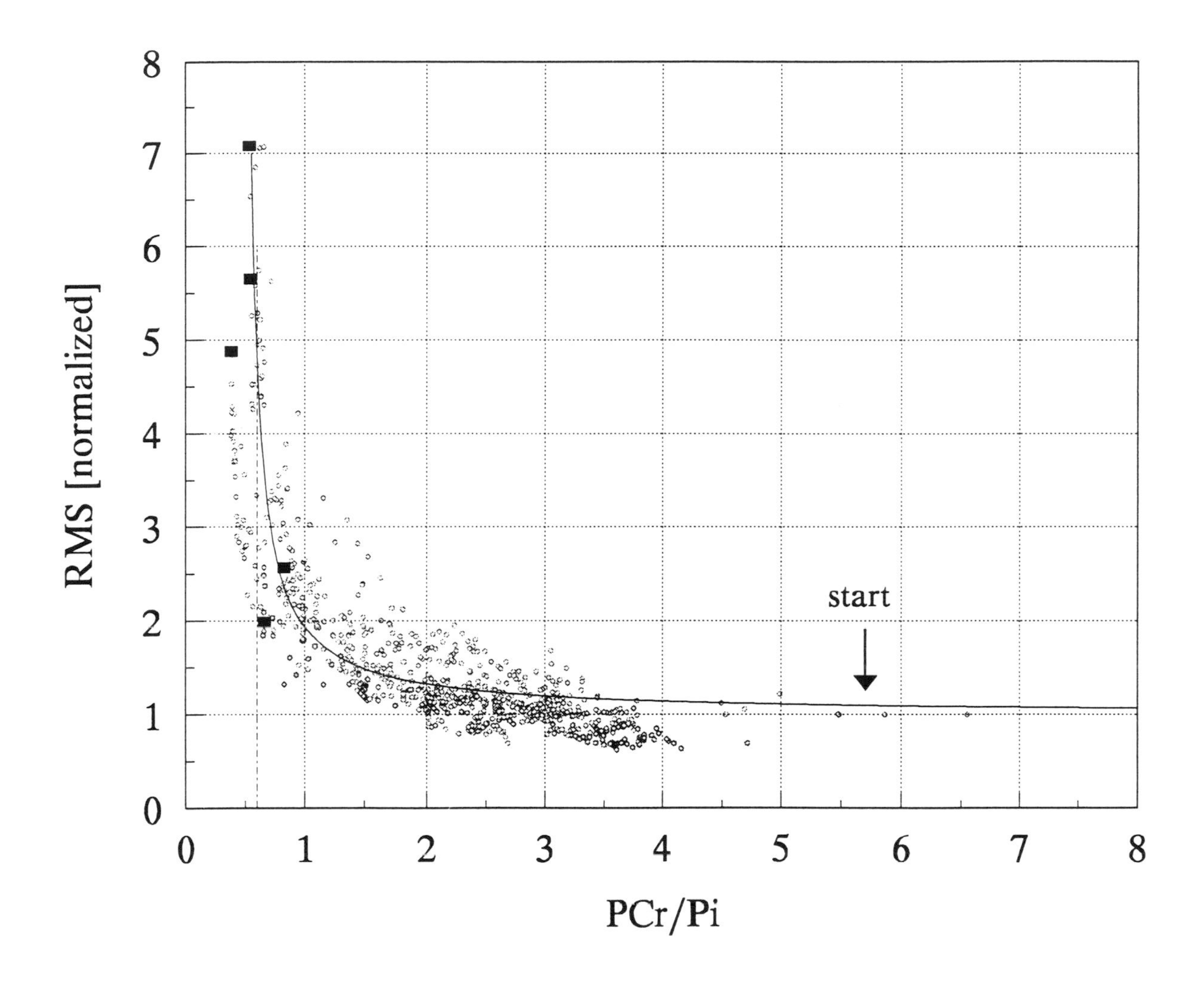

Figure 7. The RMS-PCr/P_i correlation diagram during a 10% MVC isometric contraction until exhaustion (pooled data N=5). Large filled symbols indicate the exhaustion values. The RMS values were normalized in relation to the initial value. The hyperbola: $Y=6/(1+12\cdot(X-0.55))+1$ fitted the data very well. The vertical dashed line indicates the PCr/P_i = 0.6 asymptote.

A similar metabolic limit has been reported by Chance et al. (1985), where the myocardial response to induced hypoxia was examined in a canine study: PCr/P_i ratios near 0.6 were associated with tachycardia and severe acidosis and a PCr/P_i ratio of 0.6 led to imminent heart failure.

The myoelectrical activity changes when pH is near 6.8

A concomitant RMS increase and a more rapidly decreasing MPF corresponding to a pH threshold of 6.8 followed the period of steady-state levels of PCr/P_i in MVC-10 as shown in figure 8b. This was also present during MVC-30 as well as in the exhausted subjects performing MVC-5 (III) and in the first study (I). At the threshold point the PCr/P_i levels were 1-2.

Figure 8. Correlation diagrams of a SEMG and ^{31}P NMRS during a 10% MVC isometric contraction until exhaustion and 30 min recovery. It should be noted that the fitted curve plot was based on correlation of polynomial approximation of the time series, as indicated by the sign (*). Open symbols are exercise and closed symbols are recovery. a) RMS versus Pcr/P_i, b) RMS versus pH, c) MPF versus PCr/P_i and d) MPF versus pH.

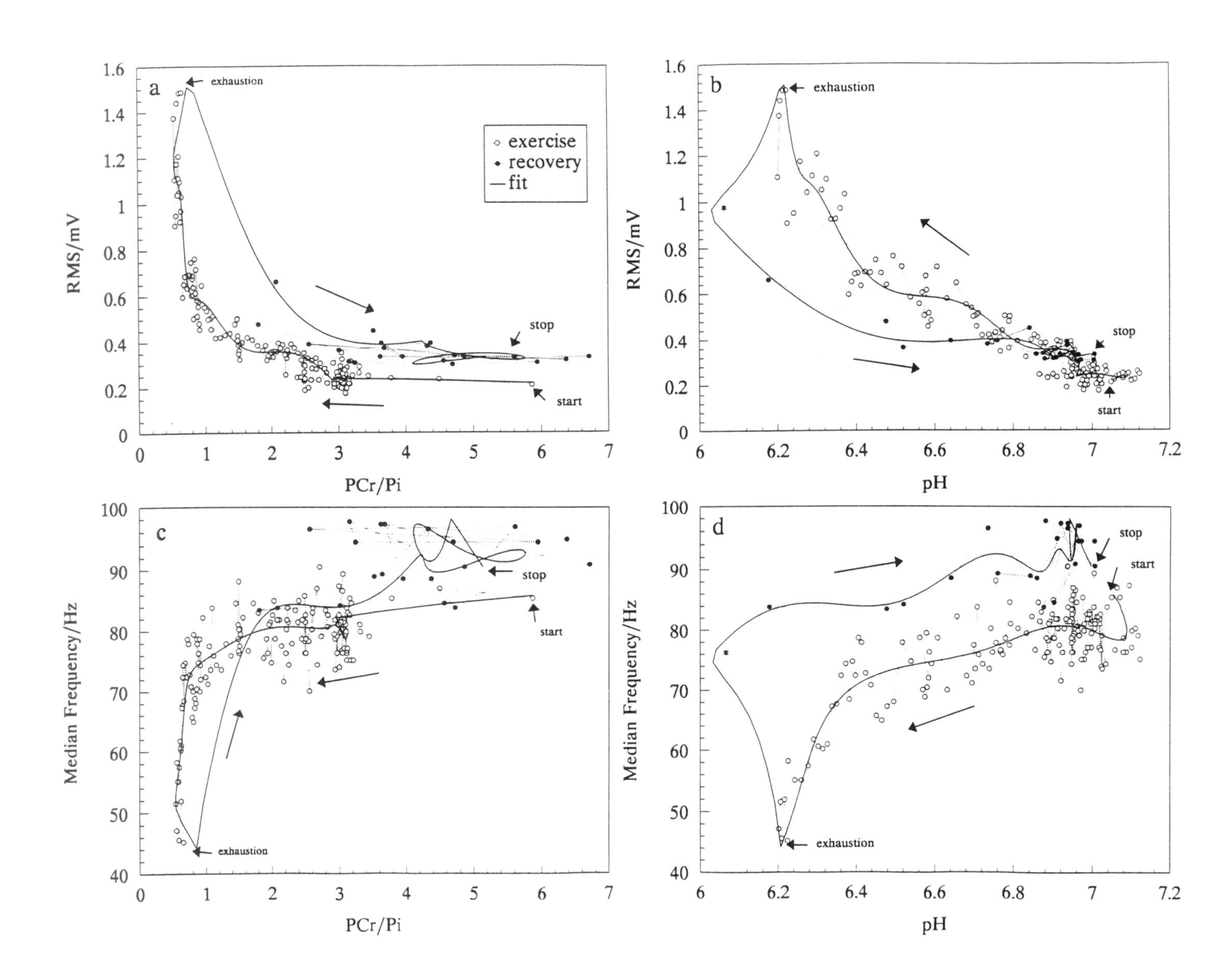

In a graded arm exercise model, a PCr/P_i ratio equal to or less than unity was associated with a beginning glycolysis and significant lactic acidosis (Chance et al. 1985) and the PCr/P_i was near 2.1 at the limit of the oxidative phase in graded exercise (Kent-Braun et al. 1993). In addition, maximum O_2 uptake has been shown to occur at pH values of 6.8 (Systrom et al. 1990). At the same point of time the P_i began to split up indicating a pH differentiation of the muscle fibre types (Park et al. 1987; Vandenborne et al. 1991; Mizuno et al. 1994). P_i splitting was not observed in the non-exhausted MVC-5 subjects. The RMS increase and MPF decrease were not observed in the non-exhausted subjects performing MVC-5, where a steady-state pH level of 6.95 and PCr/P_i near 3 were sustained throughout the exercise.

In the McArdle patient there was a hyperbolic RMS-PCr/P_i relation similar to that of the healthy volunteers. The increasing pH during exercise then indicates that the cause of the additional recruitment of motor units during exhaustive exercise is not directly initiated by proton accumulation, but happens at PCr/P_i levels near 1.5, similar to that of healthy volunteers. This PCr/P level, however, is associated with pH levels of 6.8 in healthy volunteers.

The PCr-pH relation during exercise and recovery forms a consistent hysteresis curve

The PCr-pH relation during rest, exercise and recovery exhibited a clear hysteresis curve (III) as shown in figure 9. The relation was virtually independent of contraction level and endurance. The results indicate that a higher rate of glycogenolysis is initiated when PCr has been reduced to 70 % of the resting level. This corresponded to a PCr/P_i of 1.5 which again correlated to the RMS increase at pH near 6.8. The PCr-pH recovery showed clearly that pH continued to decline and PCr recovery was more rapid than pH. The PCr-pH relation could be a standard curve showing the normal response for use in certain clinical investigations as indicated by a study of postviral fatigue syndrome (Arnold et al. 1984b). A previous study has also indicated a similar PCr-pH relation and PCr threshold, but was not shown during recovery (Taylor et al. 1983). A quasi-circular PCr-pH response during exercise and recovery was shown in (Helpern et al. 1986) using transcutaneous electrical nerve stimulation.

The recovery rate of PCr is approximately four times higher than the pH recovery rate

The PCr recovery rate during MVC-10 was not significantly different from MVC-30 with time constants less than one minute, and recovery was almost immediate in the non-exhausted MVC-5. An increasing mono-exponential curve fitted the recovery of PCr very well, and the time constants were calculated for all experiments in MVC-10 and MVC-30, but they could not be calculated for the nonexhausted MVC-5 due to the rapid recovery (II). The PCr was fully recovered after 30 min and recovery was clearly affected by the brief test contractions necessary to collect SEMG during recovery. This was more distinct the higher the contraction level, but the recovery rate does not seem to have been considerably delayed when compared to earlier reports (Taylor et al. 1983, Arnold et al. 1984a, McCully et al. 1991).

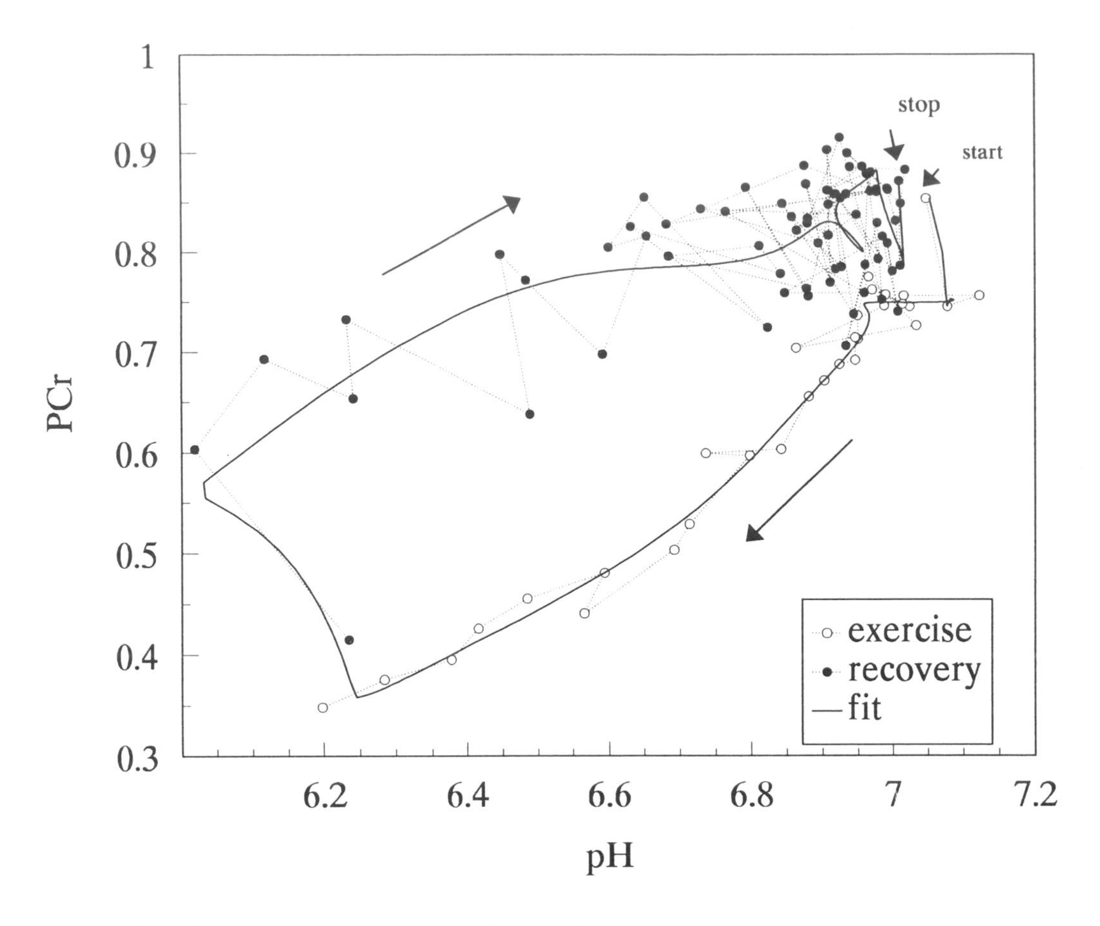

Figure 9. Correlation diagram of PCr versus pH during a 10% MVC isometric contraction until exhaustion. It should be noted that the fitted curve was based on correlation of polynomial approximation of the time series. Open symbols are exercise and closed symbols are recovery.

During the initial phase of recovery the pH continued to decline before starting to increase in a mono-exponential manner (III). The recovery rate of pH was significantly lower than the PCr, but was also fully recovered after 30 min. The PCr recovery rates were in general approximately four times higher than the pH recovery rates, which is in accordance with other studies (Arnold et al. 1984a, Pan et al. 1991; Marsh et al. 1993, Iotti et al. 1993). It has been suggested that PCr resynthesis is controlled by pH (Arnold et al. 1984a), but the study of healthy volunteers showed a poor correlation between end-pH and PCr recovery rate (III) in contrast to an earlier study (Iotti et al. 1993). This can probably be explained by the narrow end-pH range at exhaustion (III). The pH recovery was not visibly affected by the brief contractions in the recovery period.

The PCr recovery rate was significantly reduced in the McArdle patient (III), and was only the half that of the healthy volunteers. This corresponds to a lower oxidative capacity, and has also been shown in previous studies (Haller et al. 1985, Bendahan et al. 1992). After 30 min the RMS in the McArdle patient had not recovered to values comparable to values obtained at the very beginning of the contraction, but the PCr/P_i and pH were fully recovered, and could not have accounted for the unrecovered myoelectric parameters.

During recovery in the healthy volunteers the RMS rapidly decreased whereas the pH continued to decline corresponding to the time series observations (III). During the further recovery the pH increased within the first minute and RMS decreased but at a correspondingly lower level than during exercise. This persisted until pH reached levels of 6.8, from where the RMS remained at a higher level than during the beginning of exercise, and the pH recovered to resting levels (see figure 8b).

The splitting of P_i probably indicates recruitment of larger and glycolytic motor units

The splitting of the P_i resonance into two peaks with two different pH values and the RMS increase at a total pH near 6.8 might indicate that, following a temporary steady-state period of a fatiguing voluntary contraction, the already active oxidative motor units lost their force contractile properties and larger glycolytic motor units are gradually recruited according to the size principle (Henneman et al. 1965) and motor unit fibre distribution (Burke et al. 1973). The newly recruited large glycolytic motor units will dominate the SEMG and increase the RMS. The almost immediate recovery of RMS and the relatively fast recovery of PCr/P_i and PCr compared to pH could indicate that the motor units regain their force generating capacity and the oxidative potential very rapidly, and the exhausted glycolytic motor units probably account in part for the slower pH recovery. This is supported by the more rapid recovery of the P_i peak with highest pH (slow-twitch, oxidative) compared to the P_i peak with lowest pH (fast-twitch, glycolytic). In favour of this, the PCr recovery rate has recently been shown to be highly correlated to the percentage of slow-twitch fibres (Mizuno et al. 1994) also supported by earlier studies (Park et al. 1987; Vandenborne et al. 1991).

Even at exhaustion, P_i splitting was not observed in the McArdle patient, and no visible steady-state levels during the exercise were seen (III). The lack of P_i splitting at exhaustion in the McArdle patient supports the theory that a pH difference between oxidative and glycolytic fibre can be observed during exercise in healthy muscle. The missing steady-state levels and a report of impaired sympathetic activity during static exercise in McArdle patients (Pryor et al. 1990), could indicate that by-products from the breakdown of glycogen are a regulating factor in normal muscle. Supporting this, T_2 enhancement after exercise was absent in McArdle patients (Fleckenstein et al. 1991), and suggests that glycogenolysis is required in mediating the T_2 enhancement, probably due to the water redistribution that normally accompanies exercise.

The spectral changes of the SEMG described by MPF are not strongly coupled to pH or lactate

During recovery from exhaustive exercise the RMS and MPF almost immediately recovered and rapid PCr recovery was seen, but the pH continued to decline before an increase towards resting levels started, as shown in figures 8d and 8b. The reason for this, is the rapid PCr resynthesis which is a proton producing process (Taylor et al. 1983; Arnold et al. 1984a; Helpern et al. 1986). A transient pH increase during the initial rapid PCr breakdown at the beginning of exercise illustrates the opposite process, because PCr hydrolysis is a proton consuming process.

Reduced muscle fibre conduction velocity cannot solely be due to proton accumulation as suggested by several studies where the MPF has been shown to mirror the conduction velocity (Arendt-Nielsen and Mills 1985; Eberstein and Beattie 1985; Zwarts 1985), because the MPF recovery is almost immediate in contrast to the far slower recovery of pH. The simultaneous pH increase and the MPF decline during exercise in the patient with McArdle's syndrome also supports this notion (III). The MPF decline in a McArdle patient was also observed in a previous study (Mills and Edwards 1984) but the force declined and this could have contributed to the decline of MPF as shown in Broman et al. (1985). This finding is in contrast to previous studies where a clear linear relation between muscle fibre conduction velocity and median frequency of the SEMG was found during exercise. However these studies solely implicated the exercise period.

The results indicate that mechanisms other than lactate and proton accumulation may contribute to the MPF changes induced by a reduction of the propagation velocity along the muscle fibres. The smaller decrease in MPF of the McArdle patient may also indicate that the conduction velocity changes are due to a summation of mechanisms affecting the cell membrane and that lactate may contribute, as regards healthy muscle, but a major contribution by proton accumulation seems less probable.

Some studies have reported the lack of a relationship between the conduction velocity and the MPF. In a group of McArdle patients the MPF decreased normally during ischaemic exercise, but the muscle fibre conduction velocity remained unchanged (Linssen et al. 1990). A low-level isometric contraction of an arm flexor did not lower the conduction velocity, but the MPF decreased significantly (Krogh-Lund et al. 1992). Another study reported that MPF decrease is due to both decreasing conduction velocity and changes of the fundamental shape of the individual muscle fibre action potentials in the repolarization phase (Brody et al. 1991). The fact that alterations of the action potential shape during isometric contraction change with the Na^+/Ka^+ ion balance across the muscle membrane (Juel 1988, Brody et al. 1991) could account for the observed MPF changes to some extent.

The incomplete recovery of RMS and full recovery of metabolism indicate an impaired excitation-contraction coupling

All SEMG and ^{31}P NMR spectroscopy parameters had reached resting or initial levels after 30 min recovery except the RMS which initially recovered rapidly but was still incomplete after 30 min, even in the non-exhausted MVC-5. This is visible in figure 8a and 8b. Similar observations have been described in an earlier study (Miller et al. 1987). The recovery of RMS and MPF in the McArdle patient was similar to the healthy volunteers, though an even more distinct increased RMS level was seen after 30 min recovery and the MPF recovery tended to overshoot the initial levels in spite of the metabolic parameters having already reached resting levels (III).

The results indicate that the increased RMS-amplitude of the SEMG at the same isometric force level is a long term process which does not relate to phosphorous metabolism or pH, but is seemingly associated with changes in the excitation-contraction coupling. This mechanism could be low-frequency fatigue: in a previous study force recovery after low-frequency stimulation was incomplete for many hours, whereas force recovered almost completely within 30 min after high-frequency stimulus (Edwards et al. 1977).

One mechanism that is suggested to cause an alteration of the excitation-contraction coupling may be reduced Ca^{2+}-release from the sarcoplasmatic reticulum linked to reduced Ca^{2+}-sensitivity of the myofibrils (Westerblad et al. 1991). Several studies have shown that high P_i concentrations reduce the Ca^{2+}-sensitivity of the myofilaments (Brandt et al. 1982) and the Ca^{2+}-channels are possibly affected by pH. The action of P_i and pH on sarcoplasmatic Ca^{2+}-kinetics could explain the reason for muscle fatigue, but does not directly explain why increased RMS levels are associated with resting levels of pH and P_i after a long recovery period, as shown in this study.

RELATION OF METABOLISM AND MYOELECTRICAL ACTIVITY IN PATIENTS WITH FIBROMYALGIA

The maximum voluntary muscular force of fibromyalgia patients is significantly lower than sedentary controls despite equal muscle size

The MVC of the patients was significantly lower than the sedentary controls, despite equal maximum transsectional area of the anterior tibial muscle (IV). The tendency of a lowered torque/muscle area ratio in the patients might be due solely to submaximal central activation during the estimation of the maximum force, and is in part supported by a tendency to longer endurance time. However, it has been shown that physical inactivity causes a reduction of MVC that is more pronounced than the reduction in muscle area (Appel 1990) and the high occurrence of degenerative tissue found in patients (Yunus and Kalyan Raman 1989, Schrøder et al. 1993) might indicate that the low physical activity levels of fibromyalgia patients are connected with an increased amount of connective tissue which could explain the difference in MVC despite similar muscle area. Recently, significantly reduced muscle force in relation to muscle size was found in the knee extensors of fibromyalgia patients, when the degree of central activation was taken into account (V).

The myoelectrical activity is normal in relation to metabolism during exhaustive exercise but does not exhibit the same degrees of change

The changes in the SEMG and ^{31}P NMR spectroscopy parameters during exercise and recovery had in general the same characteristics as the sedentary controls (IV) and the young male volunteers (III) studied. The young male volunteers are termed "energetic controls" according to their physical activity level.

In general the correlation analysis of the patients and sedentary controls was analogous to the MVC-10 of the energetic controls, with similar working points and thresholds (IV). The RMS-PCr/P_i relationship followed the same hyperbola found in the study of healthy volunteers (III), but the exhaustion levels seemed to be clustered in two groups along the trajectory of the hyperbola (IV). When the data of the MVC-10 from energetic controls were incorporated, the exhaustion points were roughly distributed into three clusters. Evaluating the statistical differences by rank ordering the exhaustion points along the hyperbolic process showed that the exhaustion points of the patients were not sufficiently different from the sedentary controls to be judged statistically significant. In contrast, the exhaustion points of the patients were significantly less extreme than the energetic controls. The energetic controls were also significantly more extreme than the sedentary controls (IV) as shown in figure 10.

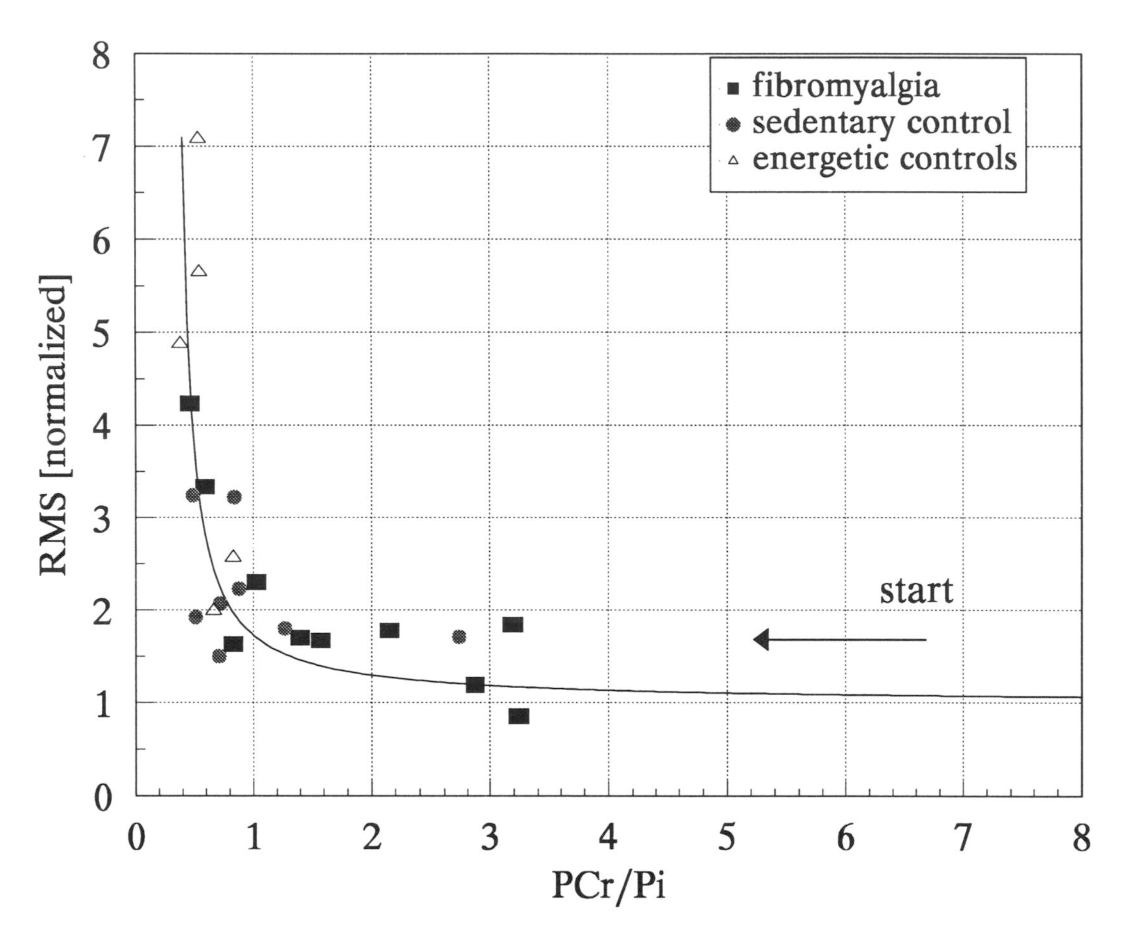

Figure 10. RMS-PCR/P_i correlation diagram of the exhaustion points after a 10% MVC isometric contraction in fibromyalgia patients, sedentary controls and energetic healthy volunteers. Large filled symbols indicate the exhaustion values. The RMS are normalized values. The hyperbola: $Y=6/(1+12\cdot(X-0.55))+1$ fitted well to the data.

The RMS-pH, MPF-PCr/P_i, MPF-pH and PCr-pH relations also indicated that the exhaustion levels in the patients seemed to be less extreme than the sedentary controls. Moreover, P_i splitting was rare, and the cause might be that only a few patients reached pH values lower than the threshold at 6.7-6.8 that was observed in the group of energetic volunteers (IV). The PCr recovery rates of the patients and the sedentary controls were significantly higher than the energetic controls, but this was probably solely due to the lower pH exhaustion values in the healthy volunteers.

The results support the theory that recovery from severe acidosis is prolonged, and that pH might be a controlling factor of PCr resynthesis. A high PCr recovery rate is normally interpreted as a high oxidative capacity (Taylor et al. 1983, McCully et al. 1991,1992), but the different levels of pH at exhaustion of the groups do not allow a true comparison.

Low physical activity levels may be connected with a lower tolerance of exercise

Although the exhaustion values of the patients were not significantly different from the sedentary controls, the tendency was that all parameters at exhaustion were less extreme than the control group (IV). Further, when compared with the energetic controls, the very considerably smaller changes in the patients and the smaller changes in the sedentary controls indicate that the exhaustion levels might be associated with the physical activity level of each group. The sedentary controls had higher daily energy consumption than the patients in spite of the effort to match their physical activity levels. The findings might be explained solely by a lower central activation during exhaustive exercise, but does not explain why a marked difference was found between the sedentary and the energetic controls. The results may also indicate that subjects with low physical activity levels have a lower tolerance of glycolytic activity during voluntary contractions than subjects with a higher physical activity level. This view is supported by a study where a large group of patients with fibromyalgia were generally aerobically unfit, and a large group of the examined patients failed to reach the anaerobic threshold (Clark et al. 1992a). Besides, the patients believed they were working at a higher than expected effort. Patients following prescribed physical training for a period improved in several aspects including reduced fatigue, pain and number of tender points (McCain et al. 1988; Clark et al. 1992b; Høydalsmo et al. 1992). Despite the indications of reduced muscle force per muscle area and that the glycolytic intolerance is compatible with subjects with low physical activity levels, contributions from pathological muscle pain cannot be excluded. Recent results suggest that tender points are caused by sympathetically activated intrafusal contraction judged by spontaneous EMG activity solely in tender points (Hubbard and Berkoff 1993), and pain could probably be a factor in the exercise intolerance.

4. CONCLUSIONS

This study describes the combination of ^{31}P NMR spectroscopy and SEMG in order to investigate how myoelectrical changes are related to energy metabolism in exercising and recovering human skeletal muscle. The conclusions of the study are:

Simultaneous and non-invasive measurements of myoelectrical activity and energy metabolism in contracting human skeletal muscle are technically possible with ^{31}P NMR spectroscopy and SEMG, in a whole-body magnet surrounded by a radio frequency shielded cabin. Reliable ^{31}P NMR spectroscopy and SEMG data can be obtained during exercise and recovery with high time-resolution and known volume localisation. This was achieved by sensitivity enhancement of the ^{31}P NMR spectroscopy, signal processing, NMR imaging and filtration of the SEMG.

The results indicate that proton or lactate accumulation is not primarily responsible for the SEMG spectral changes as previously suggested. Motor unit recruitment increases hyperbolically during submaximal exhaustive static exercise, with decreasing PCr/P_i (oxidative potential) reaching maximum at a specific oxidative potential constituting a consistent metabolic limit to a voluntary muscle contraction. The myoelectrical activity changes at a certain metabolic and pH level and seems to be related to the recruitment of large glycolytic motor units. These results stem from the anterior tibial muscle during static and constant force exercise in well-motivated and healthy volunteers.

Patients with fibromyalgia have reduced maximum muscular force in relation to muscle area. The myoelectrical-metabolic relation could not be distinguished from that of healthy volunteers, but the metabolic and myoelectrical exhaustion levels were less extreme. The exhaustion levels of the fibromyalgia patients, sedentary controls and more energetic controls corresponded with the daily physical activity levels and could indicate that low physical activity levels are connected with a lower tolerance of anaerobic exercise.

Simultaneous ^{31}P NMR spectroscopy and SEMG in combination with NMR imaging offers useful information about the energy metabolism, myoelectric activity, morphology and spatial activation of human skeletal muscle, and have the potential of being an important research tool in non-invasive studies of basal muscle physiology, sports medicine, occupational physiology as well as in clinical applications.

FUTURE VIEW

Methodologically, additional sensitivity enhancement of the ^{31}P NMR spectroscopy could be performed by proton decoupling which would improve the time resolution considerably. A supplement of other signal-processing techniques in addition to the chosen NMR signal- processing procedure could produce a method with almost no operator involvement.

Further studies are needed to verify the results of this study. The effects of physical training in both fibromyalgia patients and healthy volunteers combined with the use of external muscle stimulation to observe the central activation could add important new aspects to both exercise physiology and fibromyalgia.

REFERENCES

Appell, H.J. Muscular atrophy following immobilisation. A review. Sports Med. 10: 42-58, 1990.

Arendt-Nielsen, L. and Mills, K. R. The relationship between mean power frequency of the EMG spectrum and muscle fibre conduction velocity. Electroenceph. Clin. Neurophysiol. 60: 130-134, 1985.

Arnold, D.L., Matthews P.M. and Radda, G.K. Metabolic recovery after exercise and the assessment of mitochondrial function in vivo in human skeletal muscle by means of ^{31}P NMR. J. Magn. Reson. Med. 1: 307-315, 1984a.

Arnold, D.L., Bore, P.J., Radda, G.K., Styles, P. and Taylor, D.J. Excessive intracellular acidosis of skeletal muscle on exercise in a patient with a post-viral exhaustion/fatigue syndrome. Lancet I: 1367-1369, 1984b.

Bárány, M. and Glonek, T. Phosphorous-31 nuclear magnetic resonance of contractile systems. Methods in enzymology. 85: 624-676, 1982.
de Beer, R. Report on the quantification of eurospin MRS test signals. EEC Eurospin 25: 62-74, 1990.

de Beer, R. and Van Ormondt, D., Analysis of NMR data using time domain fitting, in "In-vivo magnetic resonance spectroscopy I: probeheads and radiofrequency pulses, spectrum analysis", ed.: P. Diehl. Springer verlag, 1992.

Bendahan, B., Confort-Gouny, S., Kozak-Ribbens, G. and Cozzone, P.J. 31-P NMR characterization of the metabolic anomalies associated with the lack of glycogen phosphorylase activity in human forearm muscle. Biochem. Biophys. Res. Com. 185: 16-21, 1992.

Bengtsson A., Henriksson K. G., Larsson J. Reduced high-energy phosphate levels in the painful muscles of patients with primary fibromyalgia. Arthritis Rheum. 29: 817-821, 1986.

Bengtsson, A. and Henriksson, K.G. The muscle in fibromyalgia--a review of Swedish studies. J. Rheumatol. Suppl. 19: 144-149, 1989.

Bernus, G., Gonzalez De Suso, J.M., Alonso, J., Martin, P.A., Prat, J.A. and Arus, C. ^{31}P-MRS of quadriceps reveals quantitative differences between sprinters and long-distance runners. Med. Sci. Sports Exerc. 25: 479-484, 1993.

Bigland-Ritchie, B. and J.J. Woods, Changes in muscle contractile properties and neural control during human muscular fatigue. Muscle & Nerve 7: 691-699, 1984.

Bigland-Ritchie, B., Cafarelli E., Vøllestad N. K. Fatigue of submaximal static contractions. Acta Physiol. Scand. 128: 137-148, 1986a.

Bigland-Ritchie, B., Furbush F., Woods J. J. Fatigue of intermittent submaximal voluntary contractions: central and peripheral factors. J. Appl. Physiol. 61: 421-429, 1986b.

de Blecourt A. C., Wolf R.F., van Rijswijk M. H., Kamman R. L., Knipping A.A., Mooyaart E. L. In vivo 31P magnetic resonance spectroscopy (MRS) of tender points in patients with primary fibromyalgia syndrome. Rheumatol. Int. 11: 51-54, 1991.

Brandt, P.W., Cox, R.N., Kawai, M. and Robinson, T. Regulation of tension in skinned muscle fibres. Effects of cross-bridge kinetics on apparent Ca^{2+} sensitivity. J. Gen. Physiol. 79: 997-1016, 1982.

Brody, L.R., Pollock, M.T., Roy, S.H., de Luca, C.J. and Celli, B. J. pH-induced effects on median frequency and conduction velocity of the myoelectrical signal. Appl. Physiol. 71(5): 1878-1885, 1991.

Broman, H., Bilotto, G. and de Luca, C.J. Myoelectric signal conduction velocity and spectral parameters: influence of force and time. J.Appl.Physiol. 58(5): 1428-1437, 1985.

Buchtal, S.D., Thoma, W.J., Taylor, J.S., Nelson, S.J. and Brown, T.R. In vivo T1 values of phosphorous metabolites in human liver and muscle determined at 1.5 T by chemical shift imaging. NMR Biomed. 2: 298-304, 1989.

Burke, R.E., Levine, D.N., Tsairis, P. and Zajac, F.E.III. Physiological types and histochemical profiles in motor units of the cat gastrocnemius. J.Physiol. 234: 723-748, 1973.

Burt, C.T., Koutcher, J., Roberts, J.T., London, R.E. and Chance, B. Magnetic resonance spectroscopy of the musculoskeletal system. Radiol. Clin. North America. 24: 321-331, 1986.

Chance, B., Eleff, S., Leigh, J.S., Sokolow, D. & Sapega, A. Mitochondrial regulation of phosphocreatine/inorganic phosphate ratios in exercising human muscle: a gated 31-P nuclear magnetic resonance. Proc. Nat. Acad. Sci. USA. 78: 6714-6718, 1981.

Chance, B., Clark, B.J., Nioka, S., Harihara Subramanian, V., Maris, J.M., Argov, Z. and Bode, H. Phosphorus nuclear magnetic resonance spectroscopy in vivo. Circulation 72(suppl IV): 103-110, 1985.

Chance, B., Younkin, D.P., Kelley, R., Bank, W.J., Berkowitz, H.D., Argov, Z., Donlon, E., Boden, B., McCully, K., Busit, N.M.R. and Kennaway, N. Magnetic resonance spectroscopy of normal and diseased muscles. Am. J. Med. Gen. 25: 659-679, 1986.

Clark, S.R., Burckhardt, C.S., Campbell, S.M. and Bennet, R.M. Physical fitness characteristics of 94 women with fibromyalgia. Scand. J. Rheum. suppl. 94. Abstr. No. 92., 1992a.

Clark, S.R., Burckhardt, C.S., Campbell, S.M. O'Reilly, C.A. and Bennet, R.M. Prescribing exercise for patients with fibromyalgia. Scand. J. Rheum. suppl. 94. Abstr. No. 93., 1992b.

Dawson, M.J., D.G. Gadian and D.R. Wilkie. Contraction and recovery of living muscles studied by ^{31}P nuclear magnetic resonance. J. Physiol. (London) 267: 703-735, 1977.

Diop, A., Briguet, A. and Graverson-Demilly, D. Automatic in vivo NMR data processing based on an enhancement procedure and linear prediction method. Magn. Reson. Med. 27: 318-328, 1992.

Eberstein, A. and Beattie, B. Simultaneous measurement of muscle conduction velocity and EMG power spectrum changes during fatigue. Muscle & Nerve 8: 768-773, 1985.

Edwards, R.H.T., Hill, D.K., Jones, D.A. and Merton, P.A. Fatigue of long duration in human skeletal muscle after exercise. J. Physiol. 272: 769-778, 1977.

Elert, J.E., Rantapaa Dahlqvist, S.B., Henriksson Larsen, K., Lorentzon, R. and Gerdle, B.U. Muscle performance, electromyography and fibre type composition in fibromyalgia and work-related myalgia, Scand. J. Rheumatol. 21: 28-34, 1992.

Ernst, R. *in* "Advances in magnetic resonance", Ed. J.S. Waugh, p.1, Academic Press 1966.

Fisher, M.J., Meyer, R.A., Adams, G.R., Foley, J.M. and Potchen, E.J. Direct relationship between proton T1 and exercise intensity in skeletal muscle MR images. Invest. Radiol. 25: 480-485, 1990.

Fleckenstein, J.L., Canby, R.C. Parkey, R.W. and Peshock, R.M. Acute effects of exercise on MR imaging of skeletal muscle in normal volunteers. AJR 151: 231-237, 1988.

Fleckenstein, J.L., Haller, R.G., Lewis, S.F., Archer, B.T., Barker, B.R., Payne, J., Parkey, R.W. and Peshock, R.M. Absence of exercise-induced MRI enhancement of skeletal muscle in McArdle's disease. J. Appl. Physiol. 71(3): 961-969, 1991.

Gadian, D. G. Nuclear magnetic resonance and its applications to living systems. Oxford University Press, 1982.

Gollnick, P.D., Piehl, K. and Saltin, B. Selective glycogen depletion pattern in human fibres after exercise of varying intensity and at varying pedalling rates. J. Appl. Physiol. 241: 45-57, 1974a.

Gollnick, P.D., Karlsson, J., Piehl, K. and Saltin, B. Selective glycogen depletion in skeletal muscle fibres of man following sustained contractions. J. Appl. Physiol. 241: 59-67, 1974b.

Häkkinen, K. and Komi, P.V. Electromyographic and mechanical characteristics of human skeletal muscle during fatigue under voluntary and reflex conditions. Electroenceph. Clin. Neurophysiol. 55: 436-444, 1983.

Haller, R.G., Lewis, S.F., Cook, J.D. and Blomqvist, C.G. Myophosphorylase deficiency impairs muscle oxidative metabolism. Ann. Neurol. 17: 196-199, 1985.

Hara,T. Evaluation of recovery from local muscle fatigue by voluntary test contractions. J. Human Erol. 9: 35-46, 1980.

Harris, R.C. Edwards, R.H.T., Hultman, E., Nordesjö, L.O., Nylind, B. and Sahlin, K. The time course of phosphorylecreatine resynthesis during recovery of the quadriceps muscle in man. Pfluegers Arch. 367: 137-142, 1976.

Helpern, J.A., Kao, W.L., Gross, B., Kensora, T.G. and Welch, K.M.A. Interleaved ^{31}P NMR with transcutaneous nerve stimulation(TNS): A method of monitoring compliance-independent skeletal muscle metabolic response to exercise. Magn. Reson. Med. 10: 50-56, 1986.

Henneman, E., Somjen, G. and Carpenter, D. Functional significance of cell size in spinal motoneurons. J.Neurophysiol. 28: 560-580, 1965.

Ho, H.K.-Y. , Snyder, R.E. and Allen, P.S. Accuracy and precision in the estimation of in vivo magnetic-resonance spectral parameters. J. Magn. Reson. 99: 590-595, 1992.

Hoult, D., Busby, S., Gadian, D., Radda, G. Richards R. and Seeley, P. Observation of tissue metabolites using ^{31}P nuclear magnetic resonance spectroscopy. Nature 252: 285-287, 1974.

Hoult, D.I., Chen, C-N, Eden, H. Elimination of baseline artifacts in spectra and their integral. J. Magn. Reson. 51: 110-117, 1983

Hubbard, D.R. and Berkoff, G.M. Myofascial trigger points show spontaneous needle EMG activity. Spine 18: 1803-1807, 1993.

Hultman, E., Del Canale, S. and Sjöholm, H. Effect of induced metabolic acidosis on intracellular pH buffer capacity and contraction force of human skeletal muscle. Clin. Sci. 69: 505-510, 1985.

Hultman, E. and I.L. Spriet. Skeletal muscle metabolism, contraction fore and glycogen utilization during prolonged electrical stimulation in humans. J. Physiol. (London) 374: 493-501, 1986.

Høydalsmo, O. Johannsen, I, Harstad, H. Jacobsen, S. and Kryger, P. Effects of a multidisciplinary training programme in fibromyalgia. Scand. J. Rheum. suppl. 94. Abstr. No. 93., 1992.

Idström, J.-P., Harihara Subramanian, V., Chance, B., Schersten, T. and Bylund-Fellenius, A.-C. Oxygen dependence of energy metabolism in contracting and recovering rat skeletal muscle. Am. J. Physiol. 248: H40-H48, 1985.

Inbar, G.F. and Noujaim, A.E. On surface EMG spectral characterization and its application to diagnostic classification. IEEE Biomed. Eng. 31: 597-604, 1984.

Iotti, S., Lodi, R., Frassinetti, C., Zaniol, P. and Barbiroli, B. In vivo assessment of mitochondrial functionality in human gastrocnemius muscle by ^{31}P MRS - The role of pH in the evaluation of phosphocreatine and inorganic phosphate recoveries from exercise. NMR Biomed. 6: 248-253, 1993.

IRPA/INIRC. International Non-ionizing Radiation Commitee of the International Radiation Protection: "Protection of the patient undergoing a magnetic resonance examination", Health Physics. 61: 923-928, 1991.

Jacobsen, S. and Danneskiold-Samsøe, B. Isometric and isokinetic muscle strength in patients with fibrositis syndrome. New characteristics for a difficult definable category of patients. Scand. J Rheumatol. 16: 61-65, 1987.

Jacobsen S., Jensen K.E., Thomsen C., Danneskiold-Samsøe B., Henriksen O. 31P magnetic resonance spectroscopy of skeletal muscle in patients with fibromyalgia. J. Rheumatol. 19: 1600-1603, 1992.

Jensen, K.E., Jacobsen, J., Thomsen, C. and Henriksen, O. Improved energy kinetics following high protein diet in McArdle's syndrome. A ^{31}P magnetic resonance spectroscopy study. Acta Neurol. Scand. 81: 499-503, 1990.

Joliot, M., Mazoyer, B.M. and Huesman, R.H. In vivo NMR spectral parameter estimation: a comparison between time and frequency domain methods. Magn. Reson. Med. 18: 358-370, 1991.

Juel, C. Muscle action potential propagation velocity changes during activity. Muscle Nerve, 11: 714-719, 1988.

Kent-Braun, J.A., Miller, R.G. and Weiner, M.W. Phases of metabolism during progressive exercise to fatigue in human skeletal muscle. J. Appl. Physiol. 75(2): 573-580, 1993.

Krogh-Lund, C. and K. Jørgensen. Modification of myo-electric power spectrum in fatigue from 15% maximal voluntary contraction of human elbow flexor muscles, to limit of endurance: reflection of conduction velocity variation and/or centrally mediated mechanisms? Eur. J. Appl. Physiol. 64: 359-379, 1992.

Kushmerick, M.J. Muscle energy metabolism, nuclear magnetic resonance spectroscopy and their potential in the study of fibromyalgia. J. Rheumatol. 16: 40-46, 1989.

Kushmerick, M.J., Moerland, T.S. and Wiseman, R.W. Mammalian skeletal muscle fibres distinguished by contents of phosphocreatine, ATP, and P_i. Proc. Natl. Acad. Sci. USA. 89: 7521-7525, 1992.

Lindström, L., Magnusson, R. and Petersen, I. Muscular fatigue and action potential conduction velocity changes studies with frequency analysis of EMG signals. Electromygraphy 4: 341-356, 1970.

Lindström, L. and Magnusson, R. Interpretation of myoelectric power spectra: a model and its application. Proc. IEEE 65: 653, 1977.

Linssen, M., Jacobs, M., Stegeman, D.F., Joosten, E.M.G. and Moleman, J. Muscle fatigue in McArdle's disease. Brain 113: 1779-1793, 1990.

De Luca, C. J. Myoelectrical Manifestations of Localizes Muscular Fatigue in Humans.
CRC Critical Reviews in Biomedical Engineering. 11(4): 250-279, 1984.

Lund, N., Bengtsson, A. and Thorborg, P. Muscle tissue oxygen pressure in primary fibromyalgia. Scand. J. Rheumatol. 15: 165-173, 1986

Luyten, P.R., Groen, J.P, Vermeulen, J.W.A.H., den Hollander, J.A. Experimental approaches to image localized human ^{31}P NMR spectroscopy. Magn. Reson. Med. 11: 1-21, 1989.

Marsh, G.D., Paterson, D.H., Potwarka, J.J. and Thompson, R.T. Transient changes in high-energy phosphates during moderate exercise. J. Appl. Physiol. 75(2): 648-656, 1993.

McArdle, B. Myopathy due to a defect in muscle glycogen breakdown. Clin. Sci. 10: 13-35, 1951.

McBroom, P., Walsh, N.E. and Dumitru, D. Electromyography in primary fibromyalgia syndrome. Clin. J. pain. 4: 117-119, 1988.

McCain G.A., Bell D.A., Mai F.M., Halliday P.D. A controlled study of the effects of a supervised cardiovascular fitness training program on the manifestations of primary fibromyalgia. Arthritis Rheum. 31: 1135-1141, 1988.

McCully, K.K., Boden, B.P., Tuchler, M. Fountain, M.R. and Chance, B. Wrist flexor muscles of elite rowers measured with magnetic resonance spectroscopy. J. Appl. Physiol. 67: 926-932, 1989.

McCully, K.K., Kakihira, H., Vandenborne, K. and Kent-Braun, J. Noninvasive measurements of activity-induced changes in muscle metabolism. J.Biomechanics 24(suppl 1): 153-161, 1991.

McCully, K.K., Candenborne, K., Demeirleir, K., Posner, J.D. and Leigh, J.S. Muscle metabolism in track athletes, using ^{31}P magnetic resonance spectroscopy. Can. J. Physiol. Pharmacol. 70: 1353-1359, 1992.

Merton, P.A. Voluntary strength and fatigue. J. Physiol. 123: 553-564, 1954.

Meyer, R.A., Brown, T.R. and Kushmerick, M.J. Phosphorous nuclear magnetic resonance of fast- and slow-twitch muscle. Am. J. Physiol. 248: C279-C287, 1985.

Miller, R.G.; Giannini, D., Milner-Brown, H.S., Layzer, R.B., Koretsky, A.P., Hooper, D. and Weiner, M.W. Effects of fatiguing exercise on high-energy phosphates, force and EMG: evidence for three phases of recovery. Muscle & Nerve 10: 810-821, 1987.

Miller, R.G., Boska, M.D., Moussavi, R.S., Carson P.J. and Weiner, M.W. ^{31}P-NMR studies of high-energy phosphates and pH in human muscle fatigue: comparison of aerobic and anaerobic exercise. J. Clin. Invest. 81: 1190-1196, 1988.

Mills, K.R. Power spectral analysis of electromyogram and compound muscle action potential during muscle fatigue and recovery. J.Physiol. 326: 401-409, 1982.

Mills, K.R. and Edwards, R.H.T. Muscle fatigue in myophosphorylase deficiency: power spectral analysis of the electromyogram. Electroenceph. Clin. Neurophysiol. 57: 330-335, 1984.

Mizuno, M., Secher, N.H. and Quistorff, B. 31-P-NMR spectroscopy, rsEMG, and histochemical fibre types of human wrist flexor muscles. J. Appl. Physiol. 76(2): 531-538, 1994.

Molé, P.A., Coulson, R.L., Caton, J.R, Nichols, B.G., Barstow, T.J. In vivo ^{31}P-NMR in human muscle: transient patterns with exercise. J. Appl. Physiol. 59: 101-104, 1985.

Moon, R.B. and Richards, J.H. Determination of intracellular pH by ^{31}P nuclear magnetic resonance. J. Biol. Chem. 284: 7276-7278, 1973.

Mortimer, J.T., Magnusson, R. and Petersén, I. Conduction velocity in ischemic muscle: effect on EMG frequency spectrum. Am. J. Physiol. 219(5): 1324-1329, 1970.

Nelson, S.J. and Brown, T.R. A method for automatic quantification of one-dimensional spectra with low signal-to-noise ratio. J. Magn. Reson. 75: 229-243, 1987.

Newham, D.J. and E.B. Cady. A ^{31}P study of fatigue and metabolism in human muscle with voluntary, intermittent contractions at different forces. NMR Biomed. 3: 211-219, 1990.

Paiss, O. and G.F. Inbar. Autoregressive modelling of surface EMG and its spectrum with application to muscle fatigue. IEEE Biomed. Eng. 34: 761-770, 1987.

Pan, J.W., Hamm, J.R., Hetherington, H.P., Rothman, D.L. and Shulman, R.G. Correlation of lactate and pH in human skeletal muscle after exercise by ^{1}H NMR. Magn. Reson. Med. 20: 57-65, 1991.

Park, J.H., Brown, R.L., Park, C.R., McCully, K., Cohn, M., Haselgrove, J. and Chance, B. Functional pools of oxidative and glycolytic fibres in human muscle observed by ^{31}P magnetic resonance spectroscopy during exercise. Proc. Natl. Acad. Sci. USA. 84: 8976-8980, 1987.

Pryor, S.L., Lewis, S.F., Haller, R.G., Bertocci, L.A. and Victor, R.G. Impairment of sympathetic activation during static exercise in patients with phosphorylase deficiency (McArdle's disease). J. Clin. Invest. 85: 1444-1452, 1990.

Quistorff, B., Nielsen, S., Thomsen, C., Jensen, K.E. and Henriksen, O. A simple calf muscle ergometer for use in a standard whole-body MR scanner. Magn. Reson. Med. 13: 444-449, 1990.

Radda, G.K. Control of bioenergetics: from cells to man by phosphorous nuclear-magnetic-resonance spectroscopy. Biochem. Soc. trans. 14: 517-525, 1984.

Rohmert, W. Untersuchung über muskelermüdung and arbeitsgestaltung. Beuth-Vertrieb, Berlin, 1962.

Ross, B.D., Radda, G.K., Gadian, D.G., Rocker, G., Esiri, M. & Falconer-Smith, J. Examination of a case of suspected McArdle's syndrome by ^{31}P nuclear magnetic resonance. New Eng. J. Med. 304(22): 1338-1343, 1981.

Roy, S.H., de Luca, C.J. and Schneider, J. Effects of electrode location on myoelectric conduction velocity and median frequency. J. Appl. Physiol. 61(4): 1510-1517, 1986.

Roy, S.H. Combined use of surface electromyography and ^{31}P-NMR spectroscopy for the study of muscle disorders. Physical Theraphy 73(12): 892-901, 1994.

Saltin, B., Sjøgaard, G. , Gaffney, F.A. and Rowell, L.B. Potassium, lactate, and water fluxes in human quadriceps muscle during static contractions. Circ. Res. 48(suppl I): 18-24, 1981.

Sapega, A., Sokolow, D.P., Graham, T.J. and Chance, B. Phosphorous nuclear magnetic resonance: a non-invasive technique for the study of muscle bioenergetics during exercise. Med. Sci. sports. exerc. 19: 410-420, 1987.

Schrøder, H.D., Drewes, A.M. and Andreasen, A. Muscle biopsy in fibromyalgia. J. Muscul. Pain. 1: 165-169, 1993.

Sjøgaard, G., Kiens, B. Jørgensen, K. and Saltin, B. Intramuscular pressure, EMG and blood flow during low-level prolonged static contraction in man. Acta Physiol. Scand. 128: 475-484, 1986.

Sjöholm, H., K. Sahlin, L. Edström and E. Hultman. Quantitative estimation of anaerobic and oxidative energy metabolism and contraction characteristics in intact human skeletal muscle in response to electrical stimulation. Clinical Physiol. 3: 227-239, 1983.

Stulen, F.B. and de Luca, C.J. Frequency Parameters of the Myoelectric Signal as a Measure og Muscle Conduction Velocity. IEEE Biomed. Eng. BME-28(7), 515-523, 1981.

Systrom, D.M., Kanarek, D.J., Kohler, S.J. and Kazemi, H. ^{31}P nuclear magnetic resonance spectroscopy study of the anaerobic threshold in humans. J. Appl. Physiol. 68(5): 2060-2066, 1990.

Taylor, D.J., Bore, P.J., Styles, P.S., Gadian, D.G. and Radda, G.D. Bioenergetics of intact human muscle a ^{31}P nuclear magnetic resonance study. Mol. Biol. Med. 1: 77-94, 1983.

Tesch, P. and Karlsson, J. Lactate in fast and slow twitch skeletal muscle fibres of man during isometric contraction. Acta Physiol. Scand. 99: 230-236, 1977.

Tesch, P.A., Komi, P.V., Jacobs, I., Karlsson, J. and Viitasalo, J.T. Influence of lactate accumulation of EMG frequency spectrum during repeated concentric contractions. Acta Physiol. Scand. 119: 61-67, 1983.

Thomsen, C. Effekten af pneumatisk tourniquet på human tværstribet skeletmuskulatur, bedømt ved 31-P-NMR-spektroskopi. Ugeskrift for læger, 30(150): 1850-1852, 1988.

Thomsen, C., Jensen, K.E. and Henriksen, O. ^{31}P NMR measurements of T2 relaxation times of metabolites in human skeletal muscle in vivo. Magn. Reson. Imaging 7: 557-559, 1989a.

Thomsen, C., Jensen, K.E. and Henriksen, O. In vivo relaxation times of ^{31}P metabolites in human skeletal muscle. Magn. Reson. Imaging 7: 231-234, 1989b.

Vandenborne, K., McCully, K., Kakihira, H., Prammer, M., Bolinger, L., Detre, J.A., De Meirleir, K., Walter, G., Chance, B. and Leigh, J.S. Metabolic heterogeneity in human calf muscle during maximal exercise. Proc. Natl. Acad. Sci. USA 88: 5714-5718, 1991.

van der Veen, J.W.C., de Beer, R. Luyten, P.R. and van Ormondt, D. Accurate quantification of in vivo ^{31}P NMR signals using the variable projection method and prior knowledge. Magn. Reson. Med. 6: 92-98, 1988.

Westerblad, H., Lee, J.A., Lännergren, J. and Allen, D.G. Cellular mechanisms of fatigue in skeletal muscle. Am. J. Physiol. 261: C195-C209, 1991.

Wolfe F., Smythe H.A., Yunus M.B., Bennett R.M., Bombardier C., Goldenberg D.L., Tugwell P., Campbell S.M., Abeles M., Clark P. The American College of Rheumatology 1990 Criteria for the Classification of Fibromyalgia. Report of the Multicenter Criteria Committee. Arthritis Rheum 33: 160-172, 1990.

Yunus, M., Masi, A.T., Calabro, J.J., Miller, K.A. and Feigenbaum, S.L. Primary fibromyalgia (fibrositis): clinical study of 50 patients with matched normal controls. Sem. Arthrit. Rheum. 11(1), 151-171, 1981.

Yunus M.B. and Kalyan Raman U.P. Muscle biopsy findings in primary fibromyalgia and other forms of nonarticular rheumatism. Rheum. Dis. Clin. North Am. 15: 115-134, 1989.

Zidar, J, Bäckman, E. Bengtsson, A. and Henriksson, K.G. Quantitative EMG and muscle tension in painful muscles in fibromyalgia. Pain 40: 249-254, 1990.

Zwarts, M.J., Van Weerden, T.W. and H.T.M. Haenen. Relationship between average muscle fibre conduction velocity and EMG power spectra during isometric contraction, recovery and applied ischemia. Eur. J. Appl. Physiol. 56: 212-216, 1987.